DESTINATION MARKETING

Creating Memorable Tourism Experiences

DESTINATION MARKETING

Creating Memorable Tourism Experiences

Rupa Rathee, PhD
Pallavi Rajain, PhD

AAP | APPLE ACADEMIC PRESS

First edition published 2023

Apple Academic Press Inc.
1265 Goldenrod Circle, NE,
Palm Bay, FL 32905 USA

760 Laurentian Drive, Unit 19,
Burlington, ON L7N 0A4, CANADA

CRC Press
6000 Broken Sound Parkway NW,
Suite 300, Boca Raton, FL 33487-2742 USA

4 Park Square, Milton Park,
Abingdon, Oxon, OX14 4RN UK

Library and Archives Canada Cataloguing in Publication

Title: Destination marketing : creating memorable tourism experiences / Rupa Rathee, PhD, Pallavi Rajain, PhD.

Names: Rathee, Rupa, author. | Rajain, Pallavi, author.

Description: First edition. | Includes bibliographical references and index.

Identifiers: Canadiana (print) 2022024443X | Canadiana (ebook) 20220244464 | ISBN 9781774910269 (hardcover) | ISBN 9781774910276 (softcover) | ISBN 9781003282501 (ebook)

Subjects: LCSH: Place marketing. | LCSH: Tourism—Marketing.

Classification: LCC G155.A1 R38 2022 | DDC 910.68/8—dc23

Library of Congress Cataloging-in-Publication Data

Names: Rathee, Rupa, author. | Rajain, Pallavi, author.

Title: Destination marketing : creating memorable tourism experiences / Rupa Rathee, Pallavi Rajain.

Description: First edition. | Palm Bay, FL : Apple Academic Press, 2023. | Includes bibliographical references and index. | Summary: "This new volume, Destination Marketing: Creating Memorable Tourism Experiences, provides a snapshot view of various aspects of destination marketing, the art of using marketing to create memorable experiences for travellers at specific destinations. The book begins with an introduction of destination marketing that discusses its origin, how it evolved into its present state, important definitions, destination marketing environments, destination value chains, consumer behavior, along with information on segmentation, targeting, and positioning for destinations. The authors cover the various tourism attractors, the most common of which include heritage tourism, agro-rural tourism, natural/scenic attractions, man-made attractions, spiritual/ religious tourism, wildlife tourism, business tourism, festivals, art and culture tourism, sports and adventure tourism, wellness and medical tourism, culinary tourism, special interest tourism, and stopover tourism. The destination marketing mix is also discussed, covering the seven P's of destination marketing. Promotional tools are included as well as destination branding methods along with various brand elements: destination names, URLs, symbols, characters, slogans, and jingles. Other chapters address destination marketing organizations (such as convention and visitor bureaus that help promote and market local attractions); performance measurement tools; the use of social media and digital marketing; tried-and-true strategies for destination marketing, such as segmenting, targeting, and positioning; the role of sustainability in destination marketing; the gap between theory and practice in destination marketing; and the future of destination marketing, with a view to advancements in technology as well as health and safety issues. This book will be valuable to faculty and students in hospitality programs as well as for researchers, tourism marketing professionals, and others interested in promoting destination tourism"-- Provided by publisher.

Identifiers: LCCN 2022023100 (print) | LCCN 2022023101 (ebook) | ISBN 9781774910269 (hardback) | ISBN 9781774910276 (paperback) | ISBN 9781003282501 (ebook)

Subjects: LCSH: Tourism--Marketing. | Place marketing.

Classification: LCC G155.A1 R38 2023 (print) | LCC G155.A1 (ebook) | DDC 910.68/8--dc23/eng/20220801

LC record available at https://lccn.loc.gov/2022023100

LC ebook record available at https://lccn.loc.gov/2022023101

ISBN: 978-1-77491-026-9 (hbk)
ISBN: 978-1-77491-027-6 (pbk)
ISBN: 978-1-00328-250-1 (ebk)

About the Authors

Rupa Rathee, PhD, is an Associate Professor in the Department of Management Studies, DCRUST, Murthal, Haryana (India). She has more than 19 years of teaching experience to her credit. Her teaching and research area include marketing and HRM. She has over 50 publications to her credit in reputed national and international journals of publishers including Emerald, Sage, and Elsevier. She has contributed eight book chapters in the books of IGI Global, McGraw Hill Education, and Palgrave Macmillan publishers. She has attended more than 30 faculty development programs at various prestigious institutes, such as IIM Bangalore, IIM Calcutta, IIM Kozhikode, MDI Gurgaon, IITM Gwalior, etc. She has contributed papers to over 50 national and international conferences. She has completed seven consultancy projects and one UGC-sponsored (University Grants Commission) major research project.

Pallavi Rajain, PhD, is an Assistant Professor at Maharaja Surajmal Institute, New Delhi. She completed her PhD from DCRUST, Murthal (India) in 2021. She received her BTech from Kurukshetra University and MBA from DCRUST, Murthal, in 2010 and 2012, respectively. She has qualified UGC NET and worked as Project Fellow on a UGC sponsored major research project on Women Entrepreneurship. She has 34 publications in national and international journals, including journals of Emerald and Sage publications.

Contents

Abbreviations

BDI	belief-desire-intention
CAM	commercial accommodation monitor
DMO	destination marketing organization
DPHH	domestic promotion and publicity
EAV	equivalent advertising value
GIT	group inclusive tour
HMD	head-mounted displays
IPA	importance-performance analysis
IT	independent traveler
LTA	local tourism administration
NTHP	National Trust for Historic Preservation
NTO	national tourism office
OTA	online travel agencies
PR	public relations
RTO	regional tourism organization
SNS	social networking site
STO	state tourism office
STP	segmentation, targeting, positioning
TAM	technology adoption model
USP	unique selling proposition
VMP	visitor media program

Acknowledgments

The authors thank the Department of Management Studies, Deenbandhu Chhotu Ram University of Science and Technology, Murthal, Sonipat, Haryana, India, where most of this research has been done. The department provided he necessary resources for the completion the research for this book.

The authors also thank the marketing researchers in this field, Berndt Schmitt and Steven Pike, who started the research on these topics more than a decade ago. The book that is presented here is highly influenced by their pathbreaking research in this area.

Special thanks to the honorable Vice Chancellor Dr. Rajender Kumar Anayath, Deenbandhu Chhotu Ram University of Science and Technology, for continually being a source of inspiration. His vision to take the university to a new era of digitization and globalization through latest trends has inspired the thoughts of many.

The authors thank the editors of Apple Academic Press for publishing this book. The editor and his team greatly helped in smoothing the editing and publishing process.

Finally, the authors thank their families and friends who supported this research process. It would not have been possible without their unwavering support to complete such a project.

—**Dr. Rupa Rathee and Dr. Pallavi Rajain**
Sonepat, 2022

Preface

This book is written due to both of the authors' interest in the field of destination experiences. The world has changed in its orientation toward selling of goods and services. As such, marketers need to focus on more than just the qualities of the products; rather they need to provide memorable experiences at the time of purchase. Destination marketing is a part of such marketing where destination or tourist places are marketed to travelers.

The main purpose of this book is to provide a snapshot of various aspects of destination marketing through a detailed analysis. For this, the content of the book is divided into eight chapters starting with an introduction and leading to latest trends in destination marketing.

Chapter 1 discusses the origin of destination marketing, how it evolved into its present state. It further provides definitions related to various aspects of destination marketing, such as destination, macrodestinations, tourism, place promotion, and destination branding. Further, the chapter deals with destination marketing environment, destination value chain, and consumer behavior. Lastly, the chapter discusses segmentation, targeting, and positioning for destinations.

Chapter 2 deals with various tourism attractors that can also be called the types or forms of tourism. The most common types include heritage tourism, agro-rural tourism, natural/scenic attractions, manmade attractions, spiritual/religious tourism, wildlife tourism, business tourism, festivals, art and culture tourism, sports and adventure tourism, wellness and medical tourism, culinary tourism, special interest tourism, and stopover tourism.

Chapter 3 uncovers the destination marketing mix. First, the chapter deals with the seven Ps for destination marketing. Next, the chapter deals with the various promotional tools for destination marketing. Further, destination branding along with various brand elements such as destination names, URLs, symbols, characters, slogans, and jingles are discussed.

Chapter 4 deals with destination marketing organizations, or DMOs as they are popularly called. The chapter deals with the development of DMOs and various related terms, such as NTO, STO, and RTO. Further, the goals and key roles and marketing planning of DMOs are discussed. Next, the various stakeholders of DMOs are elaborated on along with approaches used by them to enter a market. Lastly, the various performance measures used to evaluate the effectiveness of destination marketing are studied.

Chapter 5 deals with destination experiential marketing. Further with regard to destination experience, the destination brand experience concept has been detailed. The chapter also provides steps given by previous authors that can be used to build brand experience. Lastly, the experience co-creation model is discussed along with importance of creating memorable experiences.

Chapter 6 is about destination marketing through the use of social media. In the digital world marketing through digital media, particularly social media, is inevitable. For this the most popular sources are Facebook, Twitter, Instagram, and blogs. The chapter discusses the importance of social media in destination marketing. It also provides models for understanding how destination images are created through social media. The chapter also deals with the benefits of integrating social media with destination marketing and its use in making destinations viral. Lastly, destination marketing in a tech-savvy world is discussed.

Chapter 7 describes the various strategies related to the use of destination marketing. The strategies covered related to the segmenting, targeting, and positioning of destinations. Further, strategies related to the marketing mix are discussed. Next, certain strategies in relation to branding of destinations are mentioned. Lastly, strategies related to destination experiences DMPs and use of social media are elaborated.

Chapter 8 describes the role of sustainability in destination marketing. This chapter also discusses two types of challenges faced by destination marketing in terms of marketing and research. This has led to a gap existing between theory and practice. Lastly, the future of destination marketing has been discussed with a view to advancements in technology as well as some health and safety issues.

This book is part of the authors' larger quest for understanding experiential marketing. To do this, the authors collected books, articles, magazines, and online sources including websites to collect data related to this area. Several examples have been included within the text to make concepts more interesting.

CHAPTER 1

Understanding Destination Marketing Concepts

1.1 INTRODUCTION

Destination marketing, a concept that emerged from the broader field of tourism marketing, has been gaining momentum in the past few years. It can also be said that destination marketing mainly deals with tourism. Tourism encompasses a wide range of activities like travel, hospitality, entertainment, leisure, and much more. On searching the meaning of the word *tourism* in Britannica dictionary, one finds the definition as "the act and process of spending time away from home in pursuit of recreation, relaxation and pleasure, while making use of the commercial provision of services" (Figure 1.1).

FIGURE 1.1 Destination.

Source: thejakartapost.com.

Destination Marketing: Creating Memorable Tourism Experiences. Rupa Rathee & Pallavi Rajain (Authors)

Destination refers to "a geographical space in which a cluster of tourism resources exist, rather than a political boundary" (Pike, 2007).

Majority of the tourist activities take place at destinations. The destination is "a place at which visitors temporarily base themselves to participate in tourism related interactions and activities."

Another popular term is macro-destinations, such as the United States, which "contain thousands of micro destinations, including regions, states, cities, towns, and visitor destinations within a town" (Kotler et al., 2016).

Today, people are in search of good destinations to spend their leisure time. The stressful job environments create an urge in people to move away to different places in search of enjoyment. Thus, there was a rise in interest toward these destinations among marketers. From the marketing perspective, tourism has a demand and supply side. The *demand* side refers to customers or specifically travelers who want to satisfy their needs. In contrast, the *supply* refers to the tourism industry that satisfies the needs of the customers.

The study of destination marketing would thus be helpful to anyone who seeks to develop relationship between tourism and destination marketers. Destinations marketers are concerned with the selling of places. While tourism has been in existence since earlier times, destination marketing as a term appeared only in the 1990s.

For example, the 1999 TTRA Europe conference was themed "Tourism destination marketing—gaining the competitive edge" (Ruddy and Flanagan, 1999). Ever since, several conferences and workshops have been held on this theme. There is a journal titled *Journal of Destination Marketing and Management* published by Elsevier, which was launched in 2012. This suggests the rapidly rising interest in the field, both from the marketers as well as the academicians' point of view.

1.1.1 DEFINITIONS

The first definition of *destination marketing*, as given by Wahab, Crampon and Rothfield (1976, p. 24), states it as "the management process through which the National Tourist Organisations and/or tourist enterprises identify their selected tourists, actual and potential, communicate with them to ascertain and influence their wishes, needs, motivations, likes and dislikes, on local, regional, national and international levels, and to formulate and adapt their tourist products accordingly in view of achieving optimal tourist satisfaction thereby fulfilling their objectives" (Table 1.1).

TABLE 1.1 Texts Related to Destination Marketing.

Topic		Author(s)
Destination marketing	⇨	Wahab et al. (1976), Ashworth and Goodall (1990), Goodall and Ashworth (1990), Heath and Wall (1992), Nykiel and Jascolt (1998), Kolb (2006)
Place promotion	⇨	Ashworth and Voogd (1990), Gold and Ward (1994)
Destination branding	⇨	Morgan et al. (2002, 2004)
Conference marketing	⇨	Davidson and Rogers (2006)

Source: Authors.

Another definition by Thulisile Galelekile suggests that "destination marketing is about engaging with key players in order to drive awareness of the destination, thereby driving interest so that people visit the destination. It is about finding creative ways of communicating the destination's value proposition, therefore creating the reason for people to visit your destination."

Besides destination marketing, other terms used to describe this phenomenon are *place marketing* or *place promotion*. These terms deal with the information used by marketers for the creation of positive image of places. However, destination marketers leave out the place marketing literature as separate from that of destination marketing, even though it may at times be overlapping in nature.

Destination branding is another aspect of destination marketing, which is defined as "the process used to develop a unique identity and personality that is different from all competitive destinations" (Morrison and Anderson, 2002, p. 17).

Further, an important term for businesses is *conference marketing*, which is an "opportunity for a brand to start and own the industry conversation–a chance to gather customers, partners, employees, media, analysts and others together under one umbrella" (Johnson, n.d.).

An important aspect of destination marketing or place marketing is the creation of image of the place. The image of the destination has an impact on the behavior of tourists in terms of visit intention, attitude, intent to recommend, and perception of place. Thus, the image of the destination is crucial for decision-making. There is need to study the image of the destination in order to understand the travelers' aspirations, attitude, and behavior. A common technique used to assess the image of a destination is the importance-performance analysis (IPA). Here, IPA is a quantitative approach for measuring how people feel about certain characteristics of an issue or a thing (Martilla and James, 1977). Often, this technique is used to evaluate various attributes of recreational facilities or destinations (Hammitt et al., 1996; Oh, 2001).

1.2 DESTINATION MARKETING ENVIRONMENT

1.2.1 DESTINATION MARKETING—CRITICAL SUCCESS FACTORS

The framework for destination marketing success can be divided into two broad categories: strategic appraisal and strategic decision-making. The objective of the former is to gather information that is important for strategic decision-making. The objective of the latter is taking decisions based on specific directions and policies.

i) Strategic Appraisal

It consists of two kinds of appraisal—worldwide and nationwide. The objective of *worldwide* appraisal is to analyze national environment and strategies. This can be done with the identification of environmental impact and critical resources.

Tourism all over the world is rapidly influenced by global events, trends, and forces. The tourist destination must adapt to or overcome these environmental forces in order to be in the competition. Among the *national* influencers, normally, the researchers include socio-cultural, technological, legal, economic, ecological, and political environment (Figure 1.2).

ii) Strategic Decision-Making

It comprises of three stages, namely:

- Determination of strategic directions.
- Influencing strategic positioning.
- Outlining critical factor for success of destination marketing.

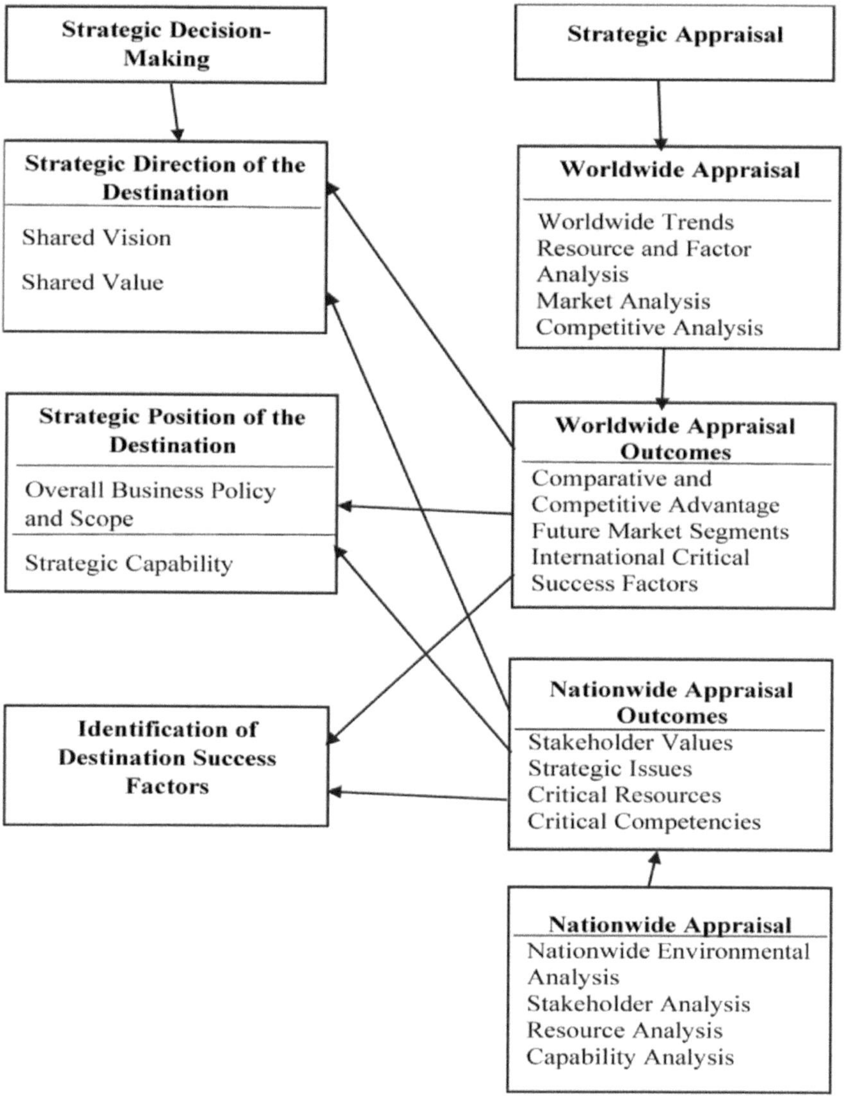

FIGURE 1.2 Destination success factors.

Source: Alhroot (2007) and Crouch and Ritchie (1999). Used with permission of the authors.

A shared value and vision integrate the outcomes of the various stakeholders. Defining a target market and positioning a destination helps to satisfy the customers.

1.2.2 PRE-CURSORS OF GROWTH FOR TRAVEL AND TOURISM AT A DESTINATION

There are four pre-cursors, which include economic, political, social, and psychological.

- *Economic*—Destinations have proven to be great source of investment as they bring in huge amounts of revenue due to large number of travelers.
- *Political*—Governments of various countries are putting a lot of effort in making policies, which are supportive of the growth of travel and tourism.
- *Social*—The rise in tourism has brought people from various communities together at a place irrespective of their religion, race, profession, personality, age, or gender.
- *Psychological*—As the world is becoming closer day by day due to ease of travel, it can be said that travel is no longer just the movement of logistics. It is the social, spiritual, and economic movement of various cultures (Mendiratta, 2014).

1.2.3 DESTINATION COMPETITIVENESS

Researchers such as Crouch and Ritchie (2000) suggested that destination competitiveness combines two aspects: the actual success of the destination and effective use of resources. The various factors related to destination competitiveness are:

a) Competitive advantage—Every destination has a vast pool of resources but it may or may not be as competitive as the destination lacking in resources. These resources comprise of historical, cultural, infrastructural, knowledge, capital, and human resources. These resources provide comparative advantage to one destination over the other.
b) Global environment—Tourism is constantly affected by global environments. Such environments may create problems and challenges or they may provide opportunities. In a world with fierce competition, such challenges or opportunities may ultimately lead to success or failure of a destination.

c) Micro environment—This is the most important factor directly dealing with the destinations. It includes destination stakeholders, travel agents, tour suppliers, and tourism markets, which provide a shape to the plans to which a destination must accommodate.
d) Policy and planning—These consist of several elements, the first of which is *defining the destination* as to what role it plays. The next is the *vision*, which reflects the philosophy of that destination. The third is the *auditing of the destination* by carrying out a critical analysis resources and functions of a destination.
e) Support factors—This includes the enterprise, facilitating resources, and infrastructure that provides the basis on which the successful foundation of a tourism industry can be laid (Line and Runyan, 2014).
f) Strategic direction—In the strategic planning phases, it is critical that a clear relationship is established between the vision and success factors. Strategic direction of a destination can be understood by understanding each stakeholder's influence (Enright and Newton, 2004).

1.3 DESTINATION VALUE CHAIN

According to Fabricius (2001), the critical support processes for destination value chain include:

1. Development of product
2. Development of infrastructure
3. Development of human resources
4. Development of systems and technology
5. Development of related industries

Firstly, for development of product, Fabricius suggests that destinations should be developed around man-made and natural attractions that satisfy the travelers' needs for accommodation, transportation, and catering.

Second, development of infrastructure requires management of security of tourists; proper navigation services, and signage on roads; aesthetic environment; improvement in public transportation like airways, railways and roadways; and amenities like water, electricity, and communications.

Third, development of human resources or people by creating jobs, improving labor relations, education and training, hospitality, and community tourism awareness. It is important to work on these aspects so that personal interaction and services are improved, which in turn enhances competitiveness.

Fourth, development of system and technology through integration of destination information system will lead to effective production, improvement in quality of services, and computerized reservations with flexible options for the convenience of the customers.

Last, the development of related industries like accommodation, recreation, transportation, entertainment, etc. The various activities of the tourism industry are critical for the success of destination marketing value chain.

1.4 CONSUMER BEHAVIOR

According to the American Marketing Association consumer behavior is the "the dynamic interaction of affect and cognition, behavior, and environmental events by which human beings conduct the exchange aspects of their lives." Further, the consumer decision process is defined as "decision-making process used by consumers regarding the market transactions before, during, and after the purchase of a good or service" (Engel et al., 1968).

1.4.1 CONSUMER DECISION-MAKING PROCESS

When a traveler is motivated, he becomes a decision-maker (Mayo and Jarvis 1981). Decisions related to where one must go, when one should go, and how to reach there must be taken. Here, the importance of destination image is reflected as it motivates one to choose a destination from a number of possible alternatives.

Consumer decision set theory is an important theory that provides possible explanation of consumer decision-making behavior. Howard (1963) and Howard and Sheth (1969) gave this theory where a total set would consist of all the available options, whether they may or may not be available and whether the traveler may or may not be aware of them.

Thus, the following possible sub-sets would be available:

- Unavailable and unaware set
- Awareness set
- Available set
- Evoked set
- Aware and unavailable set
- Available and unaware set

- Inert set
- Inept set
- Chosen destination

For a consumer to consider a destination, it is required that the destination is first of all available in the awareness set. Since a large number of destinations would be available in the awareness set, further the traveler needs to consider which of these would be more realistic options. Thus, an overlap of awareness set and available set would be taken. Then, there is the inert set, which includes options regarding which the consumer has neither a positive nor a negative opinion. The inept set consists of options that have been rejected in the initial phases. On the elimination of inert and inept destinations, the left-over destinations form the evoked set. Finally, the chosen destination comprises the destination with the greatest affinity to the purchase decision (Figure 1.3).

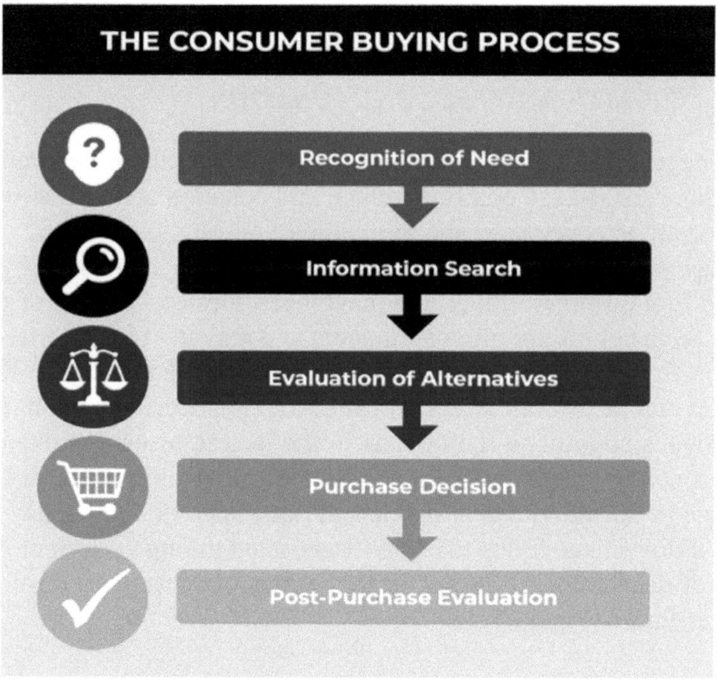

FIGURE 1.3 Consumer decision-making process.

Source: https://www.professionalacademy.com/blogs/marketing-theories-explaining-the-consumer-decision-making-process/

1.4.2 RISKS INVOLVED

There are several types of risks involved that are of concern to the tourists. These include:

Performance risk—The unpredictable weather conditions, political situations, natural hazards, crimes like theft, delays due to traffic, and airport congestion.

Social risk—Stress in social groups or social stress involved when tourists are traveling in groups. Such risk may involve embarrassing situations, individual identity representations, and travel in unfamiliar environments.

Physical risk—The transport facilities may also provide risk as the risk does not only pertain to the destination but may occur on the route.

Financial risk—Holidays involve a huge amount of financial investment, which influences the household budget. Thus, the decision to spend a significant amount of household expenditure on leisure activities involves great risk (Pike, 2007).

1.4.3 COGNITION, AFFECT, AND CONATION

The image of the destination plays an important role in dealing with all these kinds of risks. Cognition, affect, and conation are considered as the most common concepts in the formation of destination images.

Cognition generally refers to the sum of the beliefs and attitudes related to a destination that may have been formed because of previous visits. Such beliefs are often permanent and may not be influenced easily.

Affect points to the feelings about an object, which may be favorable or unfavorable. Overall affect is posited to reflect "the net resolution of an individual's cognitions (beliefs) as to the degree to which given objects possess certain attributes weighted by the salience (importance) of each attribute to the individual" (Wilkie and Pessemier 1973, p. 428).

Conative image refers to behavior reflected through action or intent. In terms of destination, it may refer to the intent of visiting a destination in the near future (Figure 1.4).

1.4.4 TOURIST MOTIVATION

Motivation may be defined as the force that guides a person's actions. Both attitude and behavior are affected due to the motivations of a person.

Tourist motivation refers to internal forces that lead a tourist to participate in tourism activities. There are five internal forces that are attached to a tourist's motivations, which include:

- Self-actualization
- Self-esteem
- Relationships
- Security
- Escapism

Further, there are certain push and pull factors that influence a tourist's motivations.

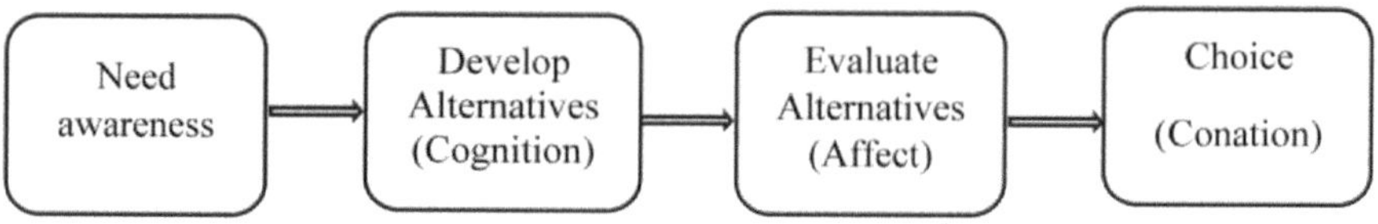

FIGURE 1.4 Cognition/affect/conation.
Source: Adapted from Myers (1992).

Push factors are related to internal demands like feelings of leisure, freedom, exploration, health, rest, and reputation.

On the other hand, *pull factors* are the visible features of a destination like beaches, hotels, entertainment, and cultural activities.

Tourist motivation influences travel decisions to a large extent. Tourist motivations can be divided into four affective levels: reputation, interpersonal, mental, and corporeal.

1.5 SEGMENTATION, TARGETING, AND POSITIONING (STP)

1.5.1 SEGMENTATION

Segmentation is an important aspect to be considered while studying destinations. It is the process of dividing a market into distinct groups of buyers who have different needs, characteristics, or behaviors and who might require separate marketing strategies or mixes.

The segmentation can be on the basis of different aspects like:

- Geographical

- Demographic
- Psychological
- Behavioral

This is mainly done so that the market may be divided into several homogenous groups. Tourist segmentation refers to "the process of assembling a group of available tourists with the same consumption demands, interests, and hobbies through marketing programs." Most of the tourism literature has focused on the demographic characteristics of tourists for segmentation.

1.5.1.1 CLASSIFICATION OF VISITOR SEGMENT

There are various ways in which visitor segments can be classified. One of these classifications is Group Inclusive Tour (GIT) and Independent Traveler (IT) depending on whether the tourists travel in a group or individually. Based on this segregation, there are certain categories of tourist segments as follows:

1. *Organized mass tourists*—They are a part of GIT. Such travelers travel in groups through tour packages provided by travel companies. They have limited contact with native people and generally remain restricted to the travel itinerary prepared by travel companies.
2. *Individual mass tourists*—This category is quite similar to the previous one but with more independence. They may travel separately to visit locations of individual's interests.
3. *Explorers*—They come under the independent traveler. They do their own planning and bookings. Generally, they have maximum interaction with the local population.
4. *Drifters*—These mainly include young travelers or, in common parlance, the backpackers. They like to mix with people of lower socioeconomic status and avoid staying in hotels.
5. *Visiting friends and relatives*—These people stay with their friends or relatives. Therefore, they are ignored by travel organizations but still they spend amount on tourist spots, shopping and eating out.
6. *Business travelers*—They spend their time on business activities like conferences, trade shows, and conventions.

7. *Pleasure travelers*—This category have a wide definition and may include various sub-classifications.
8. *Bleisure (business and pleasure travelers)*—Some business travelers like to spend time in other travel activities besides the designated business travel.
9. *Education and religious travel*—It includes travellers like students, pilgrims, missionaries and many more.
10. *Passthrough tourists*—Such visitors are of great importance to places that are located close to major tourist spots and are used as resting spots.

Another system of classification is *Plog's categorization*. In this system, the destination is categorized according to the types of visitors visiting that place. Based on this, there are two broad categories of tourists: ***allocentrics*** and ***psychocentrics***.

Allocentrics includes explorers or backpackers who visit or explore a destination that is later developed to reap economic benefits.

Psychocentrics are those people that are the least risk takers. They like to visit only familiar places that are not threatening (Kotler et al., 2016).

1.5.2 TARGETING

This step mainly includes evaluation and selection of target markets. For this, it is required that attractiveness, growth, and segment size for each segment is examined by the firm. The firms can have several options of doing target marketing: micromarketing, niche marketing, differentiated marketing, and undifferentiated marketing.

In *undifferentiated marketing* or mass marketing, the marketers ignore segmentation and target the entire market.

In *differentiated or segmented marketing*, the separate segments are targeted with separate offers.

In *concentrated or niche marketing*, the marketers target a large share of a small segment.

Lastly, *micromarketing* is making custom-made products for customers according to their specific needs and preferences (Kotler et al., 2018).

According to Chandra and Menezes (2001), there are several benefits of targeting, which include:

- Efficient allocation of resources.
- Ease of satisfying the target tourist segments.
- Developing products according to the specified tourists.

An example of targeting is Tourism Australia (National Tourism Organization of Australia). The aim of Tourism Australia was to identify the "ideal visitor segment" in order to achieve its objectives. The criteria applied by Tourism Australia included application of psychographic criteria such as lifestyle and motivations. The organization identified "experience seekers" as its ideal tourist group. Experience seekers included travelers who seek personal experiences in the form of adventure. For this, a campaign "Australian Invitation" was started in order to attract "Experience seekers." To target these customers, the destination information was even made available through National Geographic and Discovery channels (Tsiotsou and Goldsmith, 2012).

1.5.3 DESTINATION POSITIONING

Positioning means arrangement of a product to occupy a clear, distinctive, and desirable place relative to competing products in the minds of target consumers. The strategic positioning of a destination involves two important approaches: scope of business and strategic capability. The scope of business refers to where the destination is competing. The second approach deals with how to compete.

Heath and Wall (1992) define destination positioning as "the art of developing and communicating significant differences between a region or destination's tourism offering and those of challengers serving the same target markets."

There are seven stages of destination positioning:

1. Identification of target market and context for traveling.
2. Identification of a set of destinations, which are competitive.
3. Identification of motivations.
4. Assessment of strengths and weaknesses of each set of competitive destinations.
5. Identification of differentiated positioning.
6. Implementation of the selected position.
7. Assessment of the effectiveness of positioning strategy over the changing time.

When the positioning is effective, it can lead to competitive advantage. When a destination is successfully positioned into the customer's evoked decision set, then it has an advantage over the competitors. Positioning theory works on the following three assumptions:

- The current society is bombarded with information.
- Mind has a defence mechanism against this bombardment.
- Only way to enter the mind is through focused messages.

Thus, positioning strategy provides a strategic advantage and if it is not present, then it may lead to unwanted collision with stronger brands.

1.5.4 CONTEXT

Travel context is another important aspect of destination marketing. Context refers to the time of the year, geographic region for travel, and the type of trip being undertaken. For example, the context of travelers visiting a destination for honeymoon is totally different from those visiting the place for sports events. Hu and Ritchie (1993) were the first researchers to have studied the importance of context related to destination image research where attraction for a destination varied with the context.

1.6 SUMMING UP

Destination marketing is about engaging with key players in order to drive awareness of the destination, thereby driving interest so that people visit the destination. It is about finding creative ways of communicating the destination's value proposition, therefore creating the reason for people to visit a destination. An important aspect of destination marketing or place marketing is the creation of image of the place. Successful image creation through destination marketing can be divided into two broad categories: strategic appraisal and strategic decision-making. The importance of destination image is reflected as it motivates one to choose a destination from a number of possible alternatives. Further, there are four pre-cursors for the growth of travel and tourism, which include economic, political, social, and psychological. The destination value also provides various support processes for the development of destination marketing. The consumer behavior is an important aspect as consumer decision-making process ultimately leads to choose the destination. There are several types of risks involved that are of concern to the tourists. The tourists make decisions based on these risks and their motivations to visit a place. The marketers, therefore, need to segment, target, and position a destination in such a way that the tourists are satisfied and marketers have competitive advantage. The various visitor segments can be classified based on individual or group travel, or Plog's categorization.

KEYWORDS

- **destination marketing environment**
- **destination competitiveness**
- **destination value chain**
- **segmentation**
- **targeting**
- **positioning**

CHAPTER 2

Tourism Attractors

2.1 TYPES OF DESTINATION MARKETING

An attractor is "an event, activity, organization or the like that attracts people." Attractors create attention—they do not need to have anything to do with tourism. For example, it could be a cultural or entertainment event of limited duration (for example, a rock festival) or a natural phenomenon (for example, a total solar eclipse). An attractor becomes a tourist attraction after it is noticed by tourists (Mattsson et al., 2005). It is not restricted to a single type of destination as several types of destinations exist for tourists. Therefore, destination marketing or tourism attractors can be classified into the following types:

- Heritage tourism
- Agro-rural tourism
- Natural/Scenic attractions
- Man-made attractions
- Spiritual/Religious tourism
- Wildlife tourism
- Business tourism
- Festivals
- Art and culture tourism
- Sports and adventure tourism
- Wellness and medical tourism
- Culinary tourism
- Street tourism/Special interest tourism
- Stopover tourism

Destination Marketing: Creating Memorable Tourism Experiences. Rupa Rathee & Pallavi Rajain (Authors)

2.1.1 HERITAGE TOURISM

Heritage places are those "which help an understanding of the past; enrich the present and which will be of value to the future generations." Heritage tourism is defined by the National Trust for Historic Preservation (NTHP) as "travelling to experience the places, artefacts, and activities that authentically represent the stories and people of the past and present. It includes cultural, historic and natural resources" (National Trust for Historic Preservation, 2008). In heritage tourism marketing, the enriched past of the place is the main attraction besides the place or destination serving as the only product. In this type of marketing, it is very important to improve visitor's satisfaction and service quality.

The diverse culture and history all over the world are great attractors for tourists worldwide. The way the local people live is also appealing to tourists who find the diversity in the ways of life amusing. Many world organizations like UNESCO are working to preserve this cultural diversity. The differences in language, ethnicity, and race are all part of cultural differences. For example, Bhutanese culture is very much influenced by sacred Buddhist values. Hence, there is no dearth of monasteries, chortens, lhakhangs, etc., in the country, which are preserved as heritage sites. Heritage tourism marketing is not only given a boost through advertisements and publicity in mass media but social networking sites have also added to its success. Further, the marketer should try to provide an enriched experience to the tourists to gain their interest (Hasan and Jobaid, 2014). Some of the most popular examples are Colosseum in Rome, Taj Mahal in India, Pyramids at Giza, and the Great Wall of China (Figures 2.1–2.4).

2.1.2 AGRO-RURAL TOURISM

Agritourism is known to be one of the oldest forms of sustainable tourism practices with maximum contribution to economies of rural areas. Hence, it can be referred to as agro-rural tourism as well. Attracting travelers from other places to farm areas, which act as tourist attractions results in creating income as the tourists will need boarding and lodging facilities along with money spent on local stores. Agro-rural tourism also helps to protect the rural heritage and contributes to the cultural aspects. Besides the monetary benefits, agro-rural tourism also helps to create connection between consumers and the producers leading to loyal customers, which ultimately help to incur higher profits.

FIGURE 2.1 The Colosseum, Rome.
Source: Photo by FeaturedPics. https://creativecommons.org/licenses/by-sa/4.0/

FIGURE 2.2 Taj Mahal in India.
Source: © Vyacheslav Argenberg / http://www.vascoplanet.com/

FIGURE 2.3 Pyramids of Egypt.

Source: Photo by Ricardo Liberato. https://creativecommons.org/licenses/by-sa/2.0/deed.en

FIGURE 2.4 Great Wall of China.

Source: Photo by Canary Wu. https://creativecommons.org/licenses/by-sa/2.0/

Good promotion strategies are required for continuous growth of local food products. By establishing their online presence, the farmers can popularize their product among urban population as well who demand speciality products. Among the popular form of communication is social media where daily farm updates can be posted and consumers can get information about when products will be available. Another common method for communication is word of mouth, which is also considered to be the most effective method according to some survey studies (Rilla et al., 2011).

At first, these farms promoted their produce on individual basis and therefore could not create a destination-based identity for themselves. Even some farmers had thought that creating a collaborative strategy for promotion of agro-rural tourism would be more advantageous than individual efforts but effective efforts could not be taken in this direction. Therefore, it is required that synergistic efforts are taken by collaboration of agro-rural tourism so that identities of places can be created with focus on all attractions at that place (Frian, 2010).

There are several reasons due to which tourists are motivated toward agro-rural tourism. First, they may visit a site for its historical significance. Second, such sites, which are closer to cities, have several options for recreation but when they are located at a distance then they have to focus on outdoor activities like hunting, trekking, etc. Third and the most significant reason for increase in agro-rural tourism is the demand for freshly produced food products. Some authors have suggested the following reasons for agro-rural tourism: shopping, services, environment, experience, adventure, and relaxation (Che et al., 2005; Srikatanyoo and Campiranon, 2010).

Further, it can be said that all activities of agro-rural tourism contribute toward sustainability of that destination.

2.1.3 NATURAL/SCENIC ATTRACTIONS

The natural scenic attractions include those destinations where nature has a hand in creating that place beautiful. These sites have been created without human intervention with nature taking its own course. They include the fauna and flora occurring but in an unusual way, which proves to be a spectacle for its visitors like Victoria Falls, Great Barrier Reef in Australia, Dudhsagar Falls in Goa, and the Grand Canyon in Arizona (Figures 2.5–2.7).

FIGURE 2.5 Great Barrier Reef, Australia.

Source: Photo by Toby Hudson. https://creativecommons.org/licenses/by-sa/3.0/

FIGURE 2.6 Dudhsagar Falls, Goa.

Source: Photo by Srividya. https://creativecommons.org/licenses/by-sa/4.0/

FIGURE 2.7 Grand Canyon, USA.

Source: Photo by Lennart Sikkema. https://creativecommons.org/licenses/by/3.0/

Physiography includes the visual appeal or climatic conditions of a destination. For example, in summers, tourists prefer to move to locations that are comparatively cool whereas during winters it is vice versa, like mountains and beaches.

2.1.4 MAN-MADE ATTRACTIONS

Man-made destinations for tourism mean those which were made by human at different points of time and have now become tourist attractions. Several man-made structures or landmarks with interesting architecture attract tourists. For example, the Eiffel Tower in Paris, Statue of Unity in Gujarat (India), Statue of Liberty in New York, London eye and London Bridge in London, Glass Corridor with waterfall in China, and Burj Khalifa in Dubai (Figures 2.8–2.14).

FIGURE 2.8 Eiffel Tower, Paris.
Source: Photo by Benh Lieu Song. https://creativecommons.org/licenses/by-sa/3.0/

FIGURE 2.9 Statue of Unity.
Source: Photo by Vijay B. Barot. https://creativecommons.org/licenses/by-sa/4.0/

FIGURE 2.10 Burj Khalifa.

Source: Photo by Donaldytong. https://creativecommons.org/licenses/by-sa/3.0/

FIGURE 2.11 Statue of Liberty.

Source: Photo by Mcj1800. https://creativecommons.org/licenses/by-sa/4.0/

FIGURE 2.12 Zhangjiajie Grand Canyon Glass Bridge, Hunan, China

Source: Photo by HighestBridges. https://creativecommons.org/licenses/by-sa/4.0/

FIGURE 2.13 London Bridge.

Source: Photo by Mvkulkarni23. https://creativecommons.org/licenses/by-sa/3.0/

FIGURE 2.14 London Eye.

Source: Photo by Ank Kumar. https://creativecommons.org/licenses/by-sa/4.0/deed.en

2.1.5 SPIRITUAL/RELIGIOUS TOURISM

Spiritual tourism is defined as an act of traveling overseas or domestically to visit spiritual places such as temples, churches, mosques, spiritual gardens, caves, rocks, etc.

a) Temples: Famous temples include Mata Vaishno Devi, Vittala Temple, Hampi, Shri. Saibaba Samadhi Temple, and Golden Temple (Figure 2.15).

2.1.5.1 MATA VAISHNO DEVI

The holy shrine of Shri. Mata Vaishno Devi is located in the folds of three-peaked mountains called Trikuta. The visitors to the shrine called yatris undertake a trek of 12 kms from the base camp at Katra to reach the holy cave of the Mother. The cave is open for darshans throughout the year. Shri. Mata Vaishno Devi Shrine Board also known as the Shrine Board was established in 1986 to manage and regulate the yatras and makes it a satisfactory experience for the Yatris (Figure 2.16).

FIGURE 2.15 Vittala Temple, Hampi, India.

Source: Photo by Suraj S. Rao. https://creativecommons.org/licenses/by-sa/4.0/deed.en

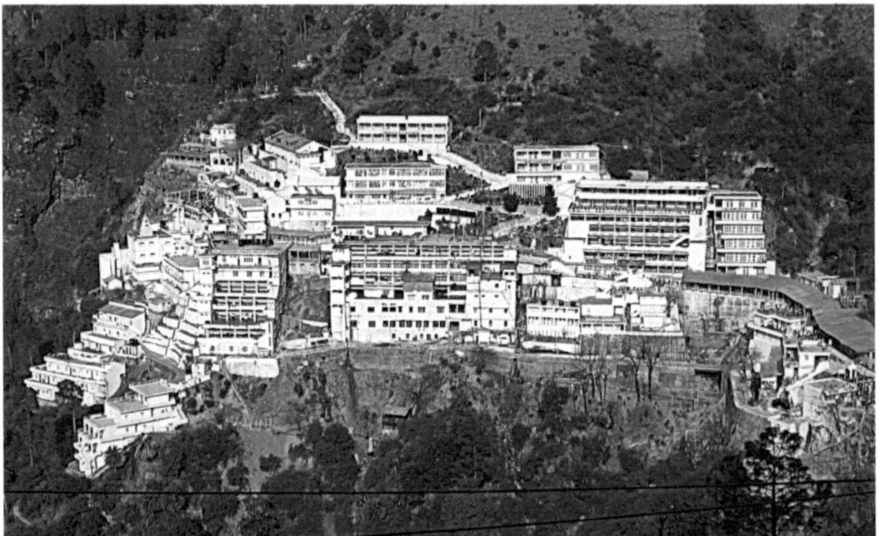

FIGURE 2.16 Vaishno Devi.

Source: Photo by Abhishek Chandra.

2.1.5.2 SHIRDI SAI DHAM

Shirdi Sai Baba was an Indian master of spirituality who was regarded as a saint by his devotees. A temple devoted to him was built in Shirdi village of Maharashtra known as Shri. Saibaba Samadhi Temple. It is a secular place where are religions are treated equal with the famous quote "Sabka Malik Ek." The temple and other temples in the premise are managed by the Shirdi Saibaba Sansthan Trust who control and provide facilities like accommodation, meals, etc. (Figures 2.17 and 2.18).

FIGURE 2.17 Sai Dham.
Source: tripadvisor.com.

2.1.5.3 GOLDEN TEMPLE

In Sikhism, The Harmandir Sahib or Golden Temple, is the most important gurdwara or temple. It was overlaid with gold leaf after initially being built from marble in the early 1600s. Inside the temple, visitors can find the Guru Granth Sahib, the holy text of Sikhism. Amrit Sarovar, or Pool of Nectar, is a body of water around the temple. An impressive site within the temple is a dining hall where volunteers serve food called langar to about 3500 people in need.

FIGURE 2.18 Golden Temple, Amritsar.
Source: Photo by Marsmux. https://creativecommons.org/licenses/by-sa/4.0/

b) Mosques: The most famous mosque is at Mecca (Figure 2.19).

FIGURE 2.19 Mecca.
Source: Photo by Tahir mq. https://creativecommons.org/licenses/by-sa/4.0/deed.en

The holiest place in Islam is Mecca where non-Muslims are not allowed. Every year, during the Hajj, millions of Muslims embark on a journey to Mecca. At least once in their lifetime, every able-bodied adult has to carry out the Hajj as it is central to Islam and is one of the religion's Five Pillars, provided it is affordable for them. Mecca is where Muhammad received the first revelation of the Qur'an as he was the prophet who founded Islam and it was his birthplace. Islam's most sacred building is the Kaaba, inside Grand Mosque where Pilgrims visit in order to praise Allah (Sood, 2012) (Figure 2.20).

c) Church: Sistine Chapel in Vatican City.

FIGURE 2.20 Vatican City.

Source: Photo by Alvesgaspar. https://creativecommons.org/licenses/by-sa/4.0/

Since 1378, Vatican City or the Holy See, has been the home of the Pope, the leader of the Catholic Church but it became an independent state in 1929. The apostle St. Peter, the first pope who was crucified and buried at St. Peter's Basilica, houses his tomb. Other attractions include the Sistine Chapel, the Vatican Museums, and the Vatican Gardens (Sood, 2012).

2.1.6 WILDLIFE TOURISM

Natural environments such as wildlife sanctuaries for birds and animals, forests, national parks, botanical gardens, lake, and oceans are part of wildlife tourism.

Some famous wildlife sanctuaries include the Gir Wildlife Sanctuary (for Lions), Periyar Wildlife Sanctuary (for elephants) (Figure 2.21), and so on.

FIGURE 2.21 Gir National Park and Wildlife Sanctuary

Source: Photo by Mayankvagadiya. https://creativecommons.org/licenses/by-sa/3.0/

Some famous national parks include the Kruger National Park in South Africa, Jim Corbett National Park (for Tigers), Kaziranga National Park (for Rhinoceros), and Kanha National Park (for Tigers) (Figures 2.22–2.25).

Some famous botanical gardens include the Singapore Botanical Gardens, Longwood Gardens, USA, and Montreal Botanical Garden (Figure 2.26).

Marine Life Park is one of the world's largest oceanarium located in Singapore (Figure 2.27).

FIGURE 2.22 Periyar wildlife sanctuary.

Source: Photo by Anand2202. https://creativecommons.org/licenses/by-sa/4.0/

FIGURE 2.23 Jim Corbett.

Source: Photo by Kaushik. https://creativecommons.org/licenses/by-sa/4.0/

FIGURE 2.24 Kaziranga National Park.

Source: Photo by Diganta Talukdar . https://creativecommons.org/licenses/by-sa/4.0/

FIGURE 2.25 Kruger National Park South Africa.

Source: South African Tourism from South Africa. https://creativecommons.org/licenses/by/2.0/deed.en

FIGURE 2.26 Montreal Botanical Garden.

Source: Photo by GarrettRock. https://creativecommons.org/licenses/by-sa/3.0/deed.en

FIGURE 2.27 Marine Life Park, Singapore.

Source: Photo by Project Manhattan.

2.1.7 BUSINESS TOURISM

This form of tourism mainly involves visits undertaken along with official duty while away from home and office. Business tourism is a focussed form of tourism involving a smaller number of people. Business tourism is divided into primary (work-related) and secondary activities (leisure). Here, the primary activities tend to be the activities of greater importance. It involves individuals as well as groups traveling together for meetings, incentives, conferences, and exhibitions (Figure 2.28).

FIGURE 2.28 Business tourism.

2.1.8 FESTIVALS

Destination marketing managers spend a lot of funds on promoting destinations. There are several festivals held in different parts of the world that promote destinations by providing adequate information about sightseeing and travel. It is part of any marketing organization's competitive policy to point out a destination's charm to the customers by such international festivals. Example, Diwali and Holi in India, Tomatina in Spain, White Knight's festival in Russia, Yee Peng and Loy Krathong festivals in Chiang Mai (Thailand), and Lantern Festival in China (Figures 2.29–2.34).

FIGURE 2.29 Diwali in India.

FIGURE 2.30 Holi festival in India.

Source: Bhrigu bayan - Own work

FIGURE 2.31 Tomatina festival in Spain.

Source: Carlesboveserral - Own work

FIGURE 2.32 White Knight's festival in Russia.

Source: Alessio Damato - I was there and I took the picture

FIGURE 2.33 Yee Peng and Loy Krathong festivals in Chiang Mai in Thailand.
Source: Takeaway - Own work

FIGURE 2.34 Lantern festival in China.
Source: Alexander Synaptic - Flickr: Gaya Street during Chinese New Year

2.1.9 ART AND CULTURE TOURISM

Tourism activities like musical performances, museums, art galleries, and several other events form part of art and culture tourism. For art lovers, such places are a great attraction. Some of the examples are Madame Tussauds Wax Museum, Louvre Museum in Paris, Jehangir Art Gallery in Mumbai, musical performances at Opera House in Sydney, Bonnaroo Music and Arts Festival, Carnevale in Venice (Italy), and Mardi Gras in New Orleans, Louisiana (Figures 2.35–2.41).

FIGURE 2.35 Madame Tussauds wax museum.
Source: Jordiferrer - Own work

2.1.10 SPORTS AND ADVENTURE TOURISM

Sports tourism refers to "travel which involves either observing or participating in a sporting event while staying apart from the tourists' usual environment" (Commonwealth of Australia, 2000). Sport tourism is a fast-growing sector of the global travel industry and equates to $7.68 billion. Broadly, it is categorized into two.

FIGURE 2.36 Louvre Museum.
Source: Benh Lieu Song (Flickr) - Own work

FIGURE 2.37 Jehangir Art Gallery, Mumbai, India.
Source: Pinakpani - Own work

The "hard" definition of sport tourism refers to the "quantity of people participating at a competitive sport event." Normally, these kinds of events are the motivation that attracts visitors to the events. Olympic Games, FIFA World Cup, F1 Grand Prix, and regional events such as NASCAR Sprint Cup Series could be described as hard sports tourism.

FIGURE 2.38 Sydney opera house.
Source: BennyG3255 - Own work

FIGURE 2.39 Bonnaroo music and arts festival.
Source: Stab At Sleep from Florida, United States - Bonnaroo What Stage

The "soft" definition of sport tourism is "when the tourist travels to participate in recreational sporting, or signing up for leisure interests" (Gammon and Robinson, 2003). Hiking, skiing, running, and canoeing can be described as soft sports tourism (360runningbarcelona.com, n.d.) (Figures 2.42–2.44).

FIGURE 2.40 Carnevale in Venice.
Source: Ionut Dragu

FIGURE 2.41 Mardi Gras in New Orleans.
Source: Flickr user radiate2357 - Photo by Flickr user radiate2357 at [1]

FIGURE 2.42 NASCAR.
Source: Zach Catanzareti Photo - restart

FIGURE 2.43 Sports tourism
Source: Tomas Castelazo - Own work

2.1.11 WELLNESS AND MEDICAL TOURISM

Medical tourism refers to "people travelling abroad to obtain medical treatment" (Horowitz et al., 2007). In the past, this usually referred to those who traveled for treatment due to unavailability of treatment in home countries that were less-developed countries to major medical centers in highly developed countries.

FIGURE 2.44 Sky diving.
Source: Skydive Andes Chile from Santiago, Chile - Fernando Vidal 23

Some of the famous wellness or medical centers include: Patanjali Yogpeeth, Haridwar—It is one of the largest yoga institutes in the world. It is Baba Ramdev's project named after Rishi Patanjali. The main purpose of the institute is to popularize the practise of yoga and ayurvedic medicine. Besides conducting research and development, the institute manufactures ayurvedic medicines for commercial sale.

Ayurvedic panchakarma, Kairali—Panchakarma, a famous detoxification process in Ayurveda has been used for centuries in Kerala. Panchkarma includes five natural methods of detoxification of the body. It helps to rejuvenate the body so that a person remains youthful (Figures 2.45 and 2.46).

2.1.12 CULINARY TOURISM

It includes marketing the local delicacies of a particular region. Some of the popular culinary destinations include Italy, Spain, Thailand, and India. Some examples of culinary tourism in India include visitors at

ChokhiDhani. ChokhiDhani Group is synonymous with Rajasthani village themed hospitality since 1990.

FIGURE 2.45 Panchakarma Ayurveda.
Source: Supraja gowda - Own work

FIGURE 2.46 Medical tourism.
Source: madrehealthcare.com.

Another example is of paranthas of Murthal dhabas in Haryana, India. Initially developed as dhabas for long-distance truck drivers, these dhabas have come up as high-end restaurants attracting culinary tourists from all over India (Figures 2.47–2.49).

Some of the famous dishes from Thailand include Tom Yum Goong at P'Aor, Tom Kha Kai at Nahm Restaurant, and Pad Thai at ThipSamai. The Spanish cuisine includes Tortilla, Churros, Paella, and Gazpacho popular in Catalonia and Valencia. The popular Italian dishes include pasta, pizza of Naples, spaghetti of Rome, risotto, and ravioli (top culinary tourism destinations, 2016).

FIGURE 2.47 Rajasthani cuisine at Chokhi Dhani in India

Source: Raveesh - https://www.flickr.com/photos/chromatic_aberration/3458557873/sizes/l/

FIGURE 2.48 Dhaba in India.

Source: tripadvisor.com.

FIGURE 2.49 Pizza of Italy.

Source: Benreis at wikivoyage shared

2.1.13 STREET TOURISM/SPECIAL INTEREST TOURISM

Last, but not the least, the popular streets and shopping centers of some of the destinations attract huge crowds from all over the world. These include Tokyo's Shinjuku neighborhood, New York's Fifth Avenue, Moscow, Milan, Oxford Street in London, Mall Road (Shimla), Los Angeles's Rodeo Drive, Champs-Elysees in Paris, and Hong Kong (travelchannel.com) (Figures 2.50–2.54).

2.1.14 STOPOVER TOURISM

This includes destinations that are only used by tourists who are passing through or traveling to someplace else. This includes places close to metropolitan cities that offer good facilities for a short stay. Common stopover destinations include Texas, Dallas, Missouri, Kansas, Singapore, and Finland.

Some airlines also provide stopover packages in collaboration with tourism boards. For example, Etihad Airways along with Abu Dhabi Tourism provides stopover package for Dubai (PATA, 2019) (Figure 2.55).

FIGURE 2.50 Champs-Elysees in Paris.

Source: Jebulon - Own work

FIGURE 2.51 Oxford Street in London.

Source: Philafrenzy - Own work

FIGURE 2.52 Kabukichō in Shinjuku, Tokyo.
Source: Wikipedia.org.

FIGURE 2.53 Rodeo Drive in Los Angeles.
Source: Wikipedia.org.

FIGURE 2.54 New York's fifth avenue.
Source: Wikipedia.org.

FIGURE 2.55 Stopover tourism at Singapore
Source: I, Sengkang

2.2 SUMMING UP

There are various forms of tourism or destination attractors. The most common types include heritage tourism, agro-rural tourism, natural/ scenic attractions, man-made attractions, spiritual/religious tourism, wildlife tourism, business tourism, festivals, art and culture tourism, sports and adventure tourism, wellness and medical tourism, culinary tourism, and special interest tourism. All these forms of tourism are unique and different from each other. They provide various forms of opportunities to tourists who like looking into the past heritage, interacting with nature and wildlife, and various forms of culture, art, and festivals. The tourists seeking medical benefits can also take dual advantage of health and tourism. The stopover tourism provides the opportunity to visit two destinations in a single visit by stopping at a tourist spot close to the main destination.

KEYWORDS

- **tourism attractors**
- **heritage tourism**
- **medical tourism**
- **culinary tourism**
- **stopover tourism**

CHAPTER 3

Destination Marketing Mix

Marketing mix is "a set of tactical marketing tools that the firm blends to produce the response it wants in the target market." Marketing mix provides customer value and helps to engage customers. Generally, the marketing mix is divided in to four variables:

- Product
- Price
- Place
- Promotion

In the case of services, there are three additional variables:

- People
- Process
- Physical evidence

3.1 SEVEN P'S FOR DESTINATION MARKETING

Among the seven P's of service marketing which include product, price, place, promotion, physical evidence, people and process; there is distinctive importance of each.

Product: Product means combination of goods and services offered to the target market. The product can be divided into five levels:

- Core product (traveling).
- Basic product (hotel reservation).
- Expected product (food and safety).
- Augmented product (sightseeing).
- Potential product (tour packages).

Destination Marketing: Creating Memorable Tourism Experiences. Rupa Rathee & Pallavi Rajain (Authors)

Thus, tourism product is a composite product combining all these ***core products*** and ***supplementary service elements.*** Supplementary elements comprise consultation, order taking, hospitality, caretaking, information, and handling exceptions. Therefore, product has become the central focus of destination marketing. The experience delivered at the destination can have the best effect on the customer and thus the product has the highest level of importance among all the elements of the marketing mix.

Price: This is the amount paid by customers to procure the product. In case of tourism, setting the price is a complex process. It includes travel (air, railways, bus, etc.) and hotel charges. Pricing depends on geographic locations as well as seasonality. Some other factors influencing price are competition, distinctiveness of destination and profit margins of travel agents or tour operators. From the customer's perspective, price plays an important role in weighing the costs incurred with the benefits gained. Destination marketers must therefore not only set prices according to target customers but they must try to minimize burdensome outlays wherever possible.

Place: The location where the collection of services is bought is referred to as the place. This includes

Direct booking: Direct booking is generally preferred due to the low transaction costs involved. However, the front desk employees may form good relationships with repeat visitors and offer them booking at lower rates so as not to spoil the relations. Similar to hotels, the restaurants also use direct bookings for most of their reservations.

Online Travel Agencies (OTA): These agencies conduct their business online with no physical facilities. There are mainly two kinds of OTAs, such as opaque and nonopaque. Opaque OTAs do not disclose the brand or name of hotels until the booking is made. The nonopaque OTAs include retail, referral models, and merchants, for example, Expedia.

Global distribution systems: These systems though originally developed by airline companies to promote their sales, are nowadays being used as product catalogs for travel agents. These computerized reservation systems provide connectivity between hotels and the GDS. The top GDS include Worldspan, Galileo, Sabre, and Amadeus. The first two of these are owned by Travelport.

Travel wholesalers and tour operators: Travel wholesalers generally target the leisure market and book hotels and airline tickets. Further, travel agents or tour operators sell these packages to tourists. The packages include accommodation, transportation, meals, and entertainment. As there is a rise in number of international resorts, tour wholesales have become powerful entities in the distribution chain. Big travel companies, such as Thomas Cook, Cox and Kings, SOTC form part of distribution channels. Thus, distribution of service is also part of the place element.

Function of distribution is to provide the products in adequate quantities whenever it is needed. Different types of *distribution channels* assist in determination of the channel policies. Physical distribution deals with reducing the cost and meeting the expectations of the customers. Maintaining a balance between the two is what provides a competitive advantage for any location. Distribution channels provide services either directly or through an agent. There are three issues related to channel decisions:

- Selection
- Distribution intensity
- Integration

The distribution can be done mainly through the airline offices, chain of hotels or tourism operators.

For example, an award-winning interactive live campaign at Košice International Airport in Slovakia was launched to promote new world destinations. Another example is of Helsinki Airport that used disruptive marketing campaign in the form of a series of videos in the style of a reality TV show (Rezcomm.com, 2018).

Promotion: Promotion means the activities undertaken to communicate the advantages or qualities of a product and lure customers into buying it. Promotion or communication of services is needed to provide information to the target audience and persuade them to use the service. Promotion includes advertising, sales promotion, publicity and personal selling. Since at times, it is difficult for customers to visualize service products as they are intangible, therefore, promotion does the task of conveying the message through physical images and metaphors.

Promotion has changed drastically with the advent of new media which include digital and social media advertising (Figures 3.1 and 3.2).

Physical evidence: Physical evidence comprises of the elements which are incorporated into a service to make it tangible and somewhat measurable. Therefore, physical evidence is same to service as packaging is for a good. It includes buildings, parking, infrastructure, a detailed guide book or brochure explaining the various details regarding the language, culture,

FIGURE 3.1 Brochure for destination.

Source: Sunshineconnelly flickr user graphia - Own work

FIGURE 3.2 Thomas Cook brochure.

Source: Mtaylor848 - Own work

and traditions of the destination being visited. The physical environment or servicescape guides the customers through service process while facilitating service delivery. Therefore, it is important to manage servicescapes with utmost care in order to have greater service productivity and better customer satisfaction (Figure 3.3).

FIGURE 3.3 Taj Hotel.

Source: Envision Studio - https://www.flickr.com/photos/186488916@N07/51297096752/

People: People are the transactional interface between the company and its customers so people deliver the service. This includes local people belonging to the place being visited and how they treat tourists. It also includes travel agent, guides, and tour operators. Lastly, the hotel and travel staff of airlines and railways. The people involved must be carefully selected and trained according to the task they need to perform. The staff must not only have the technical expertise but they should also have a positive attitude and possess good interpersonal skills.

Process: The process includes the service itself. The process comprises of following steps: providing travel information, preparing itineraries, liaison among service providers, planning the tours, ticketing and providing insurance (Ukessays, 2016). The process followed is as important as the service because a poor process would lead to low productivity and dissatisfaction. When the process is badly designed, it leads to wastage of time, ineffective service delivery and disappointment for customers. Therefore, many firms have adopted such processes which use standardized procedures, implement service quality, and automation of tasks to produce efficient systems (Wirtz *et al.*, 2018).

3.2 PROMOTIONAL TOOLS

The promotional tools or service marketing communications mix broadly includes advertising, public relations, publicity, sales promotion, personal selling, trade events, and social media.

3.2.1 ADVERTISING

In the words of Kotler *et al.* (2018) advertising is "paid nonpersonal promotion of ideas or products by an identifiable sponsor." Another definition is "advertising is a paid form of communication delivered through media from an identifiable source about an organization, product, service or idea designed to persuade the receiver to take some action now or in the future" (Belch and Belch, 2007). In order to develop any advertising campaign, it is required to identify the target customers, then to set goals which relate to sales and communication. Next is allocation of budget so that an affordable approach can be taken. Third, the decisions regarding the type of medium for sending messages and its content. Lastly, creating the advertisement and assessing the outcomes in terms of sales.

Some benefits of advertising include:

- Ability to track changes resulting from advertising.
- Ability of analyzing effectiveness in reaching target segment.
- Assessment of tourist satisfaction.
- Ability to conduct pre- and post-surveys.

Majority of the governments of various countries promote tourism in their countries through advertising the destination, also called destination advertising. For example, in India, the "Incredible India" initiative by the Ministry of Tourism. Further, the various state governments in India also promote destination in their states through advertising. For example, in India, the ads by Gujarat, Rajasthan, and Kerala Tourism. Other examples include the marketing campaign by Tourism Board of Mexico "Mexico, the place you thought you knew" (Figures 3.5 and 3.6).

FIGURE 3.4 Rajasthan Tourism.

Source: Sandhya Singh - Own work

The most common media used for advertising include television, newspapers, direct mail, posters, billboards, magazines, radio, and mobiles. The messages delivered through these mass media can be reinforced by the use of direct marketing tools, such as mailings, telemarketing, and email.

FIGURE 3.5 Gujarat Tourism.

Source: JB Kalola (patel) - Own work

FIGURE 3.6 Kerala Tourism.

Source: Shibilamol vp - Own work

3.2.2 PR AND PUBLICITY

Public Relations (PR) involves "managing communications and relationships to achieve various objectives, such as building and maintaining a positive image of the firm, handling or heading off unfavorable stories or events, and maintaining positive relationships with the media." For any destination, publicity refers to exposure, for example, in the form of "word-of-mouth." Publicity can be from many sources with which the Destination Marketing Organization (DMO) may be involved directly or indirectly. On the other hand, PR specifically focuses on creating positive image of a destination. There are examples of PR which were found to be cost-effective like use of celebrity figures as ambassadors for tourism. In India, Aamir Khan was the brand ambassador for Incredible India, and later he was replaced by Amitabh Bachhan and Priyanka Chopra.

Good PR is a factor important for the success as:

- Influence of PR is more due to higher credibility.
- Cost-effective in comparison to other methods.

Some of the major PR tools include news, special events, including conferences, speeches, and tours. Further various digital media, such as websites, blogs, YouTube, Instagram, Snapchat, Pinterest, and Twitter (Kotler *et al.*, 2018) (Figure 3.7).

FIGURE 3.7 Brand ambassadors.

3.2.3 SALES PROMOTION

Sales promotion is "special incentives or excitement-building programs that encourage consumers to purchase a particular product or service typically used in conjunction with other advertising or personal selling programs."

Broadly, there are three types of sales promotion: consumer, trade, and business promotion. There are various types of ***consumer promotion,*** such as coupons, deals, premiums, sweepstakes, samples, loyalty programs, point-of-purchase displays, rebates, and product placement.

Trade promotions mainly include providing shelf space to carry a brand, promote in advertising and pushing the product to consumers. Shelf space is so scarce that manufacturers have to offer price offs, allowances, buy-backs or free goods to the retailers to get their product on the shelf. Certain tools of consumer promotion can also be used for trade promotions, such as contests, premiums, and displays.

Business promotions have multiple objectives, such as generating business leads, stimulating purchases, rewarding customers, and motivating salespeople. The major tools used for business promotions include sales contests, conventions, and trade shows. Trade shows and conventions are places to showcase products whereas sales contests are usually conducted to provide cash prize and other gifts in order to motivate good performers (Figure 3.8).

FIGURE 3.8 Sales promotion for an amusement park.

Source: Aadams - On a visit to the park in 2018

Use of sales promotion:

- For short-term objectives.
- Increase sales during weak demand periods.
- Specific to a time period, price, and customer group.

3.2.4 PERSONAL SELLING

Personal selling is the two-way flow of communication between a buyer or buyer and seller, designed to influence the buyer's purchase decision. It can be said to be the most stimulating and meaningful form of promotion for customers. This can be in the form of customer travel exhibitions like Chelsea flower show in London, Business Travel Expo in Hong Kong, Holiday and Travel Expo in Sydney. Tourism exchange meetings are generally held between tourism suppliers and retailers as well as wholesalers (Figures 3.9 and 3.10).

FIGURE 3.9 Chelsea Flower Show in London.

Source: IXIA at English Wikipedia - Transferred from en.wikipedia to Commons by Premeditated Chaos using Commons Helper.

FIGURE 3.10 Business Expo in Hong Kong.

Source: https://commons.wikimedia.org/wiki/User:LUONGYAIU_BOROKZ

Value addition through personal selling

- Information and advice provided by salespeople.
- Time-saving and simplification of buying time by salespeople.
- Relationship building by salespeople.
- Process of Personal Selling

1. *Generating and Qualifying leads*

   This step involves generating a list of potential customers (leads) and assessment of their potential (qualify). This can be done in various ways, such as trade shows, conferences, meetings, cold calls, and social media.

2. *Pre-approach*

   This step takes place before meeting the customer. Although the salesperson has gained some initial knowledge about the customer in the qualifying stage, but in this step, additional knowledge must be gained regarding the customer before meeting personally.

3. *Sales presentation*

   After obtaining enough information about the customer, the salesperson is ready for a face-to-face meeting. In this step, the beginning is the most important part as it is where the salesperson establishes the position of the customer.

4. *Close of sale*

   This step means commitment from the customer for making the purchase. This part can be stressful for some salesperson as he/she can remain emptyhanded without a successful sale.

5. *Follow-up*

   In a relationship selling, the sale is not really over as the attitude developed by the customer after the sale decides the future prospects (Grewal and Levy, 2017).

A low-cost alternative to personal selling is telemarketing. In this, a telephone is used to reach the potential customers. Although customers get frustrated due to the intrusion because of telemarketing. However, they are given the option of adding their number to "Do not call registry" if they are not interested.

3.2.5 TRADE EVENTS

Events can be said to be "planned performances or acts" according to ancient history. But in recent times, with the availability of disposable incomes, they have taken many forms. Events can be defined as "temporary and purposive gatherings of people" (Bladen, Kennell, Abson & Wilde, 2012). According to Getz (2007), event is: "a one-time or infrequently occurring event outside the normal program or activities of the sponsoring or organizing body" and "an opportunity for an experience outside the normal range of choices or beyond everyday experience."

In the business marketplace, trade shows or events is a form of publicity. Sometimes, such events garner huge media attention where a larger number of businesses can showcase their offerings. This can be done in the form of samples, exhibits, and demonstrations to make the customers aware (Wirtz *et al.*, 2018) (Figure 3.11).

The trade events take the form of event tourism when done on a large scale. **Event tourism** is a combination of event management and tourism includes marketing as well as event planning for destination marketing targets (Getz, 2008). Event tourism exists as an intersection between event management and tourism management serving as a tool for the development of a destination.

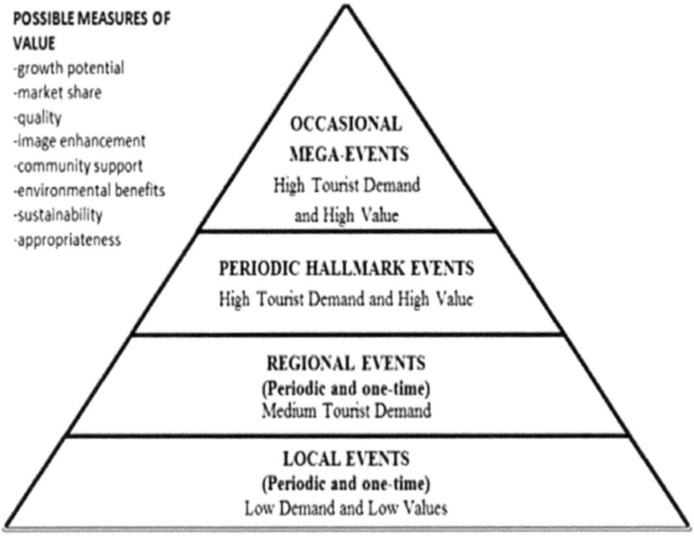

FIGURE 3.11 Portfolio approach to event tourism.

Source: Reprinted with permission from Getz (2005). © Elsevier.

Trade events generally include fairs, trade shows, conventions, meetings. For example, auto shows are events which include gatherings of people, such as manufacturers, business buyers, and public. The world's largest tourism trade fair is the ITB Berlin (Internationale Tourismus-Börse Berlin). The companies represented at the fair include car rental companies, airlines, system providers, tour operators, tourist boards, and hotels (Coban, 2016) (Figure 3.12).

In India, the Trade Fair held annually during the month of November at Pragati Maidan, Delhi is an example of international trade event (Figures 3.13 and 3.14).

FIGURE 3.12 Auto expo in India.
Source: Indian - Own work

FIGURE 3.13 Largest trade show.
Source: JATAWTF - Own work

FIGURE 3.14 Trade fair at Pragati Maidan, India.

Source: Rameshng at Malayalam Wikipedia - Transferred from ml.wikipedia to Commons by Sreejithk2000 using CommonsHelper.

3.2.6 SOCIAL MEDIA

With the advancement in technology, tourists are no longer just dependent on DMOs for destination information. They have access to various social media platforms which are a rich source of opinions. Through social media, tourists can communicate with tourism companies as well other domains which are beyond the control of tourism providers. In order to leverage the power of social media, destination marketers need to have an understanding of how and why tourists use social media. To utilize the potential of social media, tourism providers need to continuously change and upgrade themselves (Alizadeh and Isa, 2015). The usage of social media for destination marketing has been elaborated in detail in subsequent chapters (Figure 3.15).

3.3 DESTINATION BRANDING

One of the most distinct skills of professional marketers is their ability to develop and manage brands. A brand is "a name, term, sign, symbol, design or a combination of these that identifies the maker or seller of a product or service" (Kotler *et al.*, 2018). American Marketing Association defines brand as "a name, term, sign, symbol, design, or a combination of them, intended to identify the goods or service of one seller or group of sellers and to differentiate them from those of competitors." Nowadays, branding is so important that there is hardly anything that is not branded. Branding is useful for the customers also as it helps them to identify the product as well as its quality.

FIGURE 3.15 Use of social media.
Source: Doda Maria - Own work

In the case of services, the intangibility makes the evaluation of service quality difficult for customers. But with effective branding, a service brand can become synonymous with assurance and quality. Branding helps a customer to like a product but nowadays brand is evolving to the status of being loved by using the emotional connection with customers.

In the words of Leonard Barry "Strong brands enable customers to better visualize and understand intangible products. They reduce customers' perceived monetary, social, or safety risk in busying services, which are difficult to evaluate prior to purchase. Strong brands are surrogates when the company offers no fabric to touch, no trousers to try on, no watermelons or apples to scrutinize, no automobile to test drive." Thus, destination branding is a process used to develop a unique identity and personality that is different from all competitive destinations.

3.3.1 BRAND ELEMENTS

There are multiple brand elements which include brand names, URLs, logos and symbols, characters, slogans, and jingles.

a) Destination names

As suggested by Aaker (1991) "the core of any brand is its product name." An apt product name can position the brand in an effective manner by being self-explanatory. Unlike choosing names for new product being developed, destinations have already been assigned a name for centuries and several associations developed for these names. It is very difficult to change the name of places just for the sake of tourism. Although there have been examples of changing destination's names in order to create appeal for customers, for example, Hog Island was changed to Paradise Island, Elton to Surfers' paradise, Cuervo Nation an Island in the Caribbean.

3.3.1.1 INDIVIDUAL BRAND NAME

Hospitality companies can choose individual brand names for different types of products. For example, Hyatt Hotels Corporation is a multispeciality hospitality company franchises vacation properties, resorts, and luxury hotels. It uses different names, such as Grand Hyatt for large luxury hotels, Hyatt Regency for upscale brand and Hyatt Ziva for family resorts (Figure 3.16).

3.3.1.2 UMBRELLA BRAND

In this case, hospitality companies use same brand name for their entire range of products which come under the umbrella brand. For example, Marriott and Hilton that have been using umbrella branding for long. This type of branding helps in transferring the corporate image of innovativeness and trust leading to positive evaluations by customers.

b) URLs (Uniform Resource Locators)

These are specific locations on pages on the Internet which are generally referred to as domain names. With the race among companies to gather space on the web, there has been a dramatic rise in the number of registered URLs. Brand recall is the most important criteria for a URL as it increases the chances that the customers reach the website of the service. In most of the cases, the URLs are literally the brand names. For example, www.thomascook.in,www.makemytrip.com and www.goibibo.com.

FIGURE 3.16 Egypt Tourism logo
Source: Ricardo Liberato - All Gizah Pyramids

c) Destination symbols

A symbol or logo is usually designed to reflect the nature of the brand. This symbol helps to identify the brand and its personality. In the case of destinations, symbols represent well-developed icons, for example, Eiffel Tower, Pyramids, Taj Mahal, Great Wall of China, and London Bridge for many other places these may be logos. Such logos may be like kangaroos for Australia (Figures 3.17 and 3.18).

FIGURE 3.17 Australia Tourism logo.

Source: Photo by Rosino - https://www.flickr.com/photos/rosino/4778079675/, https://creativecommons.org/licenses/by-sa/2.0/

d) Characters

Characters represent a special type of brand symbol with real-life characteristics. These characters are generally introduced through advertisements and have a pivotal role in ad campaigns. These characters may be realistic or animated. For example, Mickey Mouse is a brand character for Disneyland (Figure 3.19).

e) Destination slogans

In the words of Keller (2013) "A slogan is a short phrase that communicates descriptive or persuasive information about a brand." The best example of this is the "I ♥New York" that was highly promoted during the 1970s. Many social and public organizations have held several competitions for the development of effective slogans. Some other popular slogans include "Come say "Nihau"! ...and discover the glory of China," "Incredible India," "Singapore Roars," "Uganda – the pearl of Africa," "What happens in Vegas, stays in Vegas" and "Virginia is for Lovers" (Figures 3.20 and 3.21).

f) Jingles

These are the musical messages written around the brand. They are written by professional songwriters and can be in the form of musical slogans that are extended. They are very valuable form for transmission of messages as they generally repeat the name of brand several times in interesting ways. For example, the jingle by Rajasthan tourism

"*Maatibaandhepainjanee,*
Bangdipehnebaadli,
Dedodedobaavdo,
Ghod-mathodbaavdi..."

Meaning "The land is wearing anklets, the sky is wearing bangles, the clouds in a happy frenzy have taken different shapes, the beautiful revolving and rotund earth..." (Chohan, 2016).

FIGURE 3.18 Mickey Mouse.

FIGURE 3.19 India slogan.

FIGURE 3.20 Singapore slogan.

3.3.2 LEVERAGING BRANDS

An existing brand can be leveraged by employing the following:

- *Co-branding:* It is also known as dual branding. In this type of branding, the complementary strengths of two brands can be taken advantage of by the company. For example, same retail space used by different restaurants like Taco Bell, KFC, and Pizza Hut.
- *Ingredient branding:* It can be called a special case of co-branding which provides differentiation. For example, Westin Hotels advertise their blankets, pillows, sheets and bed as "Heavenly" which has made the brand so popular that the hotel now sells these products via their online catalog.

3.4 SUMMING UP

The seven P's of service marketing or service marketing mix which include product, price, place, promotion, physical evidence, people, and process have distinctive importance for destination marketing. Among these, promotion is of special importance to destination marketers. The promotional tools or service marketing communications mix broadly includes advertising, public relations, publicity, sales promotion, personal selling, trade events, and social media. Next, the destination branding along with some important brand elements includes destination names including individual and umbrella brands, URLs, symbols, slogans, characters, and jingles. There are two methods of leveraging a brand which include co-branding or ingredient branding.

KEYWORDS

- **destination marketing mix**
- **promotional tools**
- **brand elements**
- **destination branding**

CHAPTER 4

Destination Marketing Organizations

4.1 INTRODUCTION

Destination Marketing Organizations (DMOs) mainly deal with "selling of places." This field has gained attention only in the recent years even though tourism has existed since the early 19th century, if only talking about it in an organized manner. The limited literature related to destination management, marketing, and planning has surfaced only since the 1990s.

As discussed in Chapter 1, destinations are places where travelers come to stay temporarily for various purposes. WTO (2002) offered the following definition of *tourism destinations*:

"A local tourism destination is a physical space in which a visitor spends at least one overnight. It includes tourism products, such as support services and attractions, and tourism resources within one day's return travel time. It has physical and administrative boundaries defining its management, and images and perceptions defining its market competitiveness. Local destinations incorporate various stakeholders often including a host community, and can nest and network to form larger destinations."

Wahab et al. (1976, p. 24) gave the following definition of destination marketing, "The management process through which the National Tourist Organizations and/or tourist enterprises identify their selected tourists, actual and potential, communicate with them to ascertain and influence their wishes, needs, motivations, likes and dislikes, on local, regional, national and international levels, and to formulate and adapt their tourist products accordingly in view of achieving optimal tourist satisfaction thereby fulfilling their objectives."

Destination Marketing: Creating Memorable Tourism Experiences. Rupa Rathee & Pallavi Rajain (Authors)

In the words of Inkson and Kolb (1998), organizations may be defined as "formal entities in which a complex interaction of people, materials, and money is used for the creation and distribution of goods and services."

Thus, DMO can be defined as "Any organization, at any level, which is responsible for the marketing of an identifiable destination. This therefore excludes separate government departments that are responsible for planning and policy" (Pike, 2004). However, in many countries, the marketing for destinations is done with the help of governments including local, state, and national governments, making it difficult to separate both. Therefore, the definition by Atorough & Martin seems more fit that "a DMO is a form of alliance that involves the coming together of stakeholders within an area or region, for the purpose of promoting and marketing the destinations image and attractions to potential visitors."

Goals of DMOs

- destination image enhancement
- profitability of the industry
- reduction in seasonality
- funding for long term

Further, some other terms related to DMOs include the following:

1. National Tourism Office (NTO) "The entity with overall responsibility for marketing a country as a tourism destination."
2. State Tourism Office (STO) "The organization with overall responsibility for marketing a state, province or territory as a tourism destination, in a country that has a federal political system."
3. Regional Tourism Organization (RTO) "The organization responsible for marketing a concentrated tourism area as a tourism destination."
4. Local Tourism Administration (LTA) "local government authority or a form of cooperative association of tourism businesses."

4.2 DEVELOPMENT OF DMO

The first National Tourism Office was developed almost a century ago in 1901 by the name of New Zealand Department of Tourists and Health Resorts (NZTPD, 1976). Besides this, New Zealand was also the first country to develop government tourist bureau comprising of visitor information center, wholesaler, and travel agent. New Zealand's NTO underwent several changes and restructurings in the past century, the most prominent being the one in 1991 when its governance was handed over to private sector and it was renamed New Zealand Tourism Board.

4.2.1 RISE IN EUROPE

Among the European countries, France developed the NTO in as early as 1910 while other European countries developed it 1919 onwards. In Canada, there was no National Tourism Authority until 1934 and surprisingly, USA, one of the most popular destinations in the world developed it only in 1961.

4.2.2 DEVELOPMENT IN INDIA

In India, the Ministry of Tourism was established in 1967 which was created after several deliberations subsequent to the formation of the India Tourism Development Corporation. Later, in 2002, the Incredible India campaign was launched to promote international tourism. The Ministry of Tourism further launched Domestic Promotion and Publicity (DPHH) in 2004 aimed at creating awareness with respect to potential tourist destinations. The initial success of this scheme paved the way for future investments. Further two schemes "Atithi Devo Bhava" and "Indian for India" were also launched to increase awareness.

4.2.3 GOVERNMENT INTERVENTION

Although private sector enterprises have contributed a lot to the development of the tourism, still the rapid growth would not have been possible without government intervention. Government provided a sense of security, stimulus for leisure and development of infrastructure. In the USA, during the 1980s, there were 150 government programs which directly influenced tourism. With the globalization of tourism, there has been rising competition with governments spending more on tourism ultimately leading to higher taxes. This has been resisted by residents as they do not regard tourism to be of the same importance as other government services like security, education, and health.

4.3 KEY ROLES OF DMOS

- ***Marketing***

Marketing of destinations is the key role of DMOs as suggested by their name. In order to attract travelers, the DMOs need to focus on various techniques as there is a huge competition among destinations. One destination is easily replaceable by the closest counterpart. In order to create the right image and special place in the mind of the consumers, it is needed that the destination marketers focus on appropriate marketing techniques.

- ***Coordination***

One of the major challenges for DMOs is to develop a coordinated approach among the stakeholders for beating the competition. For the DMO leaders, the greatest task is to find a common ground so that a consistent message can be conveyed. However, as seen in practice, the various stakeholders are not willing to accept a common approach. Therefore, it is the task of DMOs to find an optimal solution for a centralized approach. DMOs coordinate and cooperate with them for following reasons:

- Increase the budget for promotion.
- Providing wholesome experience to the travelers.
- Create brand synergy.
- Monitoring quality

Maintenance of quality standards is what keeps a destination apart from others in creating a unique image. The travelers would be highly disappointed if they are not provided services according the quality expected by them. Thus, it the responsibility of the DMOs to keep a check on the quality standards. Some researchers have also stressed that consistent quality standard is one of the critical success factors for DMOs.

- ***Enhancement of relations***

It is a common misconception that the primary objective of DMOs is to cater to the local tourism operators. However, from the government's perspective, the real objective of the DMOs is economic growth by providing more jobs. Therefore, DMOs are usually recommended to inform about this strategy through messages spread using specific terms that focus on economic prosperity and growth through healthy relations with all stakeholders.

4.4 MARKETING PLANNING

In tourism, there is an approach of outward–inward thinking, which suggests that the products are developed only after anticipating the tourist's needs. A DMO has to work in such a way that resident's culture is not adversely affected, needs of the suppliers are fulfilled and travelers are provided maximum benefit. In line with this, the various steps in destination marketing planning are as follows:

- A situation analysis.
- incorporating a resource audit.
- Environmental analysis and competitive positioning analysis.
- Development of goals and strategy.
- An action plan and measures of performance.

4.5 VARIOUS STAKEHOLDERS IN DESTINATION MARKETING

4.5.1 TOURISTS

According to the customer-centric approach of market orientation, the role of customers is primary. From destination marketing's perspective, the tourists are the customers or buyers. Matching the needs of the tourists according to the destination is what leads to value creation. Here, in particular independent tourists are the stakeholders as they represent nongroup tourists who choose a destination according to their satisfaction. In order to attract independent tourists, DMOs need to maintain the image of the destination according to the type of destination and the satisfaction it can lead to the tourists. The destination's image must be projected in such a way to the tourists that it can be communicated to both the current and potential tourists (Figure 4.1).

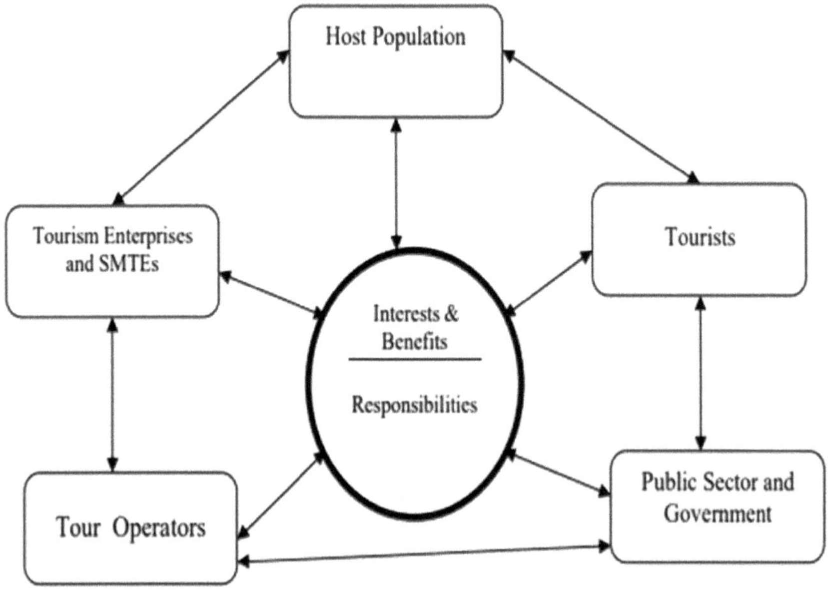

FIGURE 4.1 Stakeholders of destination marketing.
Source: Reprinted from Buhalis, 2000 with permission.

4.5.2 COMPETITORS

Competitors are as important to the marketing of destination as they are to the marketing of goods and services. Generally, destinations are similar to one another, in terms of their functional attributes and are thus very easily replaceable. In order to gain competitive advantage, the competitors must project the destination's image in a superior manner. As retailers compete for small and large volume clients, similarly, destinations compete for individual and group level tourism. Destinations cannot afford to take a short-sighted approach and ignore such threats.

4.5.3 COMMUNITY

This comprises of the politicians, local residents, and leaders of the community, etc. as they largely impact the destination marketing activities and its scope. Therefore, DMOs stay in touch with the local social and political issues. Even though decisions regarding the marketing of a destination are taken by the politicians, but still they are representatives of the community. As the local community directly impacts the activities of DMOs, they must consider the community's needs in order to create value for themselves.

4.5.4 LOCAL TOURISM

The local tourism often relies on DMOs for the creation of demand. Thus, private business owners who can gain from destination marketing are the fourth stakeholders of this market. The coordination and relationship among these private businesses and DMO determines the level of success achieved and achievement of goals. The development of a mutually beneficial relationship depends on the amount of social inclusion as part of the marketing strategy (Figure 4.2).

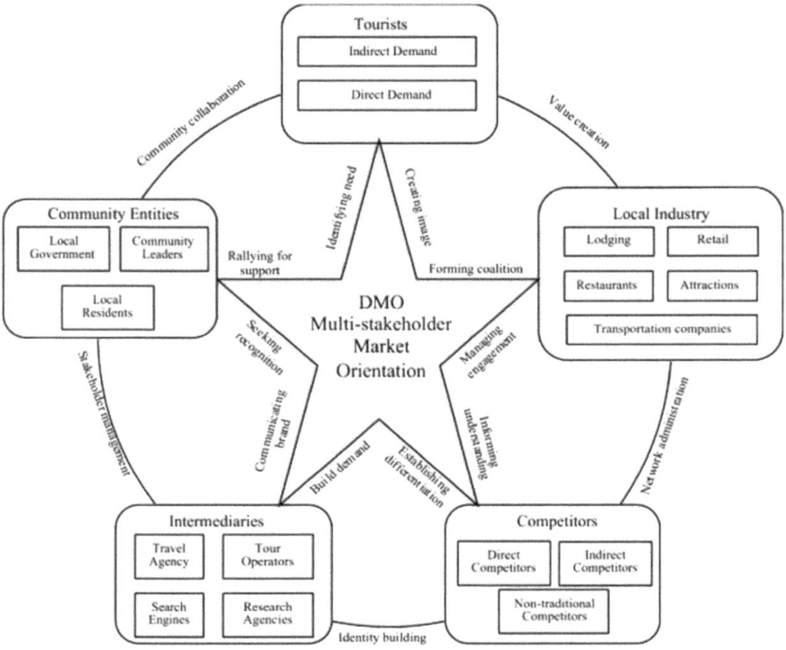

FIGURE 4.2 Stakeholders of destination marketing.

Source: Reprinted with permission from Line and Wang (2017).

4.5.5 INTERMEDIARIES

Numerous trade and research organizations that are an extension of the local tourism industry form a part of the intermediaries who are the fifth stakeholders. Although these intermediaries can take on several roles but their primary role is the creation of demand for the destination.

It is a challenge for destination marketers to align the interests of all the stakeholders.

4.6 APPROACHES USED BY DMOS TO ENTER A MARKET

Two common approaches used by DMOs include leveraging physical and financial assets, leveraging knowledge and skills.

4.6.1 LEVERAGING PHYSICAL AND FINANCIAL ASSETS

The physical assets are composed of infrastructure which may be provided by the local community at reduced prices. Additional funds can be acquired from the community for large events. These events provide high returns on investment. So, in return, the DMOs provide rebates on various services like hotels, car rentals, sightseeing.

4.6.2 LEVERAGING KNOWLEDGE AND SKILLS

The knowledge and skills under DMOs can be categorized into "networking," "marketing and sales." Networking with influential people of the community provides a good opportunity for making large booking for an event. Marketing and sales activities include participating in trade show events, publishing advertisements, providing improved services for group events. Advertisements in magazines target a specific audience for example sports fans in case of sports tourism (Pouder et al., 2018).

4.7 PERFORMANCE MEASURES

The two broad categories of performance evaluation include market performance and organization performance.

4.7.1 MARKET PERFORMANCE

Performance is a topic which has received attention only recently after 1997 (Sheehan & Ritchie; Faulkner). Until the 1990s, most of the tourism organization did not have proper effective performance measures. The organizations used some faulty performance indications with correlations that were not true indicators of the performance of promotions (Figure 4.3).

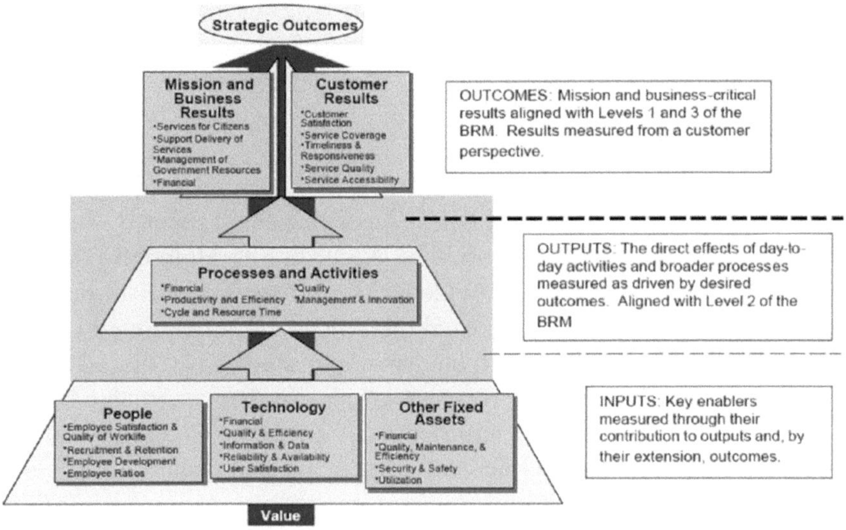

FIGURE 4.3 Performance measures.

Source: FEA - FEA Consolidated Reference Model Document

This section deals with the first category of performance measures, that is, market performance. There are further three subcategories of market performance, such as visitor monitors, marcom evaluation, and consumer-based brand equity (CBBE).

4.7.1.1 VISITOR MONITORS

The most common and obvious measure of success of a particular destination is the number of tourist arrivals. However, the collection of visitor arrival data is not an easy task. Even if data for international tourist arrivals can be obtained from immigration officials, still the job of STOs and RTOs remains difficult. The rise of visitor monitors was possible only after the emergence of DMOs. A country which has developed standards in providing tourism statistics is New Zealand. It has developed the Commercial Accommodation Monitor (CAM) which requires participation from all commercial accommodation providers. There are several detractors of visitor monitors who argue that these monitors rely too much on accommodation providers' information (as filled in forms) and question the accuracy of the data provided by them. Another form of visitor monitor is the visitor survey. It is a form of intercept survey where the data are collected from visitors of a town in detail. The survey form may include items like origin of visitors, their activities, length of stay, information needs, and accommodation.

4.7.1.2 MARCOM EVALUATIONS

Advertising: It has been said that if more money is spent by marketers, they get a greater number of visitors. According to Holcolm (1999), the "I ♥ New York" promotion provided 12% increase in tourism receipts in comparison to the previous year. Similar promotion was done in India by "I Love Chandigarh." Research on conversion studies is required to assess whether advertising leads to increase in sales, as conversion studies keep an estimate of how many enquiries from advertisements are converted into visitors. For example, the cost-effectiveness of print media could be assessed by cost per thousand readers in case of circulation figures or cost per enquiry in terms of response (Figure 4.4).

FIGURE 4.4 Advertising.

Source: Johnscotaus - Own work

Publicity: Equivalent advertising value (EAV) is an easy method to monitor Visitor Media Program (VMP) for DMOs. Equivalent advertising value is the "measure of the amount of advertising dollars required to purchase the equivalent amount of air time or column centimeter generated by the PR initiative." EAVs have proved to be useful tools for the development of public relations and enhancement of credibility among stakeholders. However, it has certain disadvantages:

- There is no clear indication of who actually views the screenings.
- The saying that "any publicity is good publicity" does not hold true.
- EAV figures do not only indicate media articles attributable to DMOs.
- Time lag between media visitation and subsequent publication.

Trade events: As monitoring trade events proved difficult so, most of the DMOs have been neglecting it. To assess the impact of travel shows on actual travel is a difficult thing to measure. It is even more difficult to separate genuine enquiries from brochure collectors at events and to anticipate those with a desire to visit a destination. Although, some travel exhibitions have started to charge fees, which has to some extent lead to the screening of prospective customers. Some DMOs even distribute coupons and for these, the redemption rate can be used as a measure. Trade fairs have ticketing system. Other example includes boat race held during Onam in Kerala, kite flying competition in Bengaluru (Figures 4.5 and 4.6).

FIGURE 4.5 Boat race in Kerala.
Source: Wikipedia.

FIGURE 4.6 Kite flying in Bengaluru.
Source: Pradeep Kumar Chatte.

4.7.1.3 CONSUMER-BASED BRAND EQUITY

Customer-based brand equity is the "differential effect that brand knowledge has on consumer response to the marketing of that brand" (Keller, 2013). The basic premise of the CBBE concept is that the power of a brand lies in what customers have learned, felt, seen, and heard about the brand as a result of their experiences over time. According to the model of Consumer-Based Brand Equity developed by Aaker (1991) which comprises of brand awareness, brand loyalty, brand associations, and perceived value. A brand which has no value to customers will not be of value to its investors.

4.7.1.4 DESTINATION AWARENESS

Generally, measuring the number of destinations inside a customer's awareness set is not a fruitful activity. What really is needed is the top-of-mind awareness and behavioral intentions. For example, if a tourist is asked that if he/she is on a short break and he has to decide a destination, what comes first to his/her mind. This prompts the tourist to limit his/her options. Thus, it can be said that brand awareness comprises brand recognition and brand recall (Keller, 2013).

4.7.1.5 DESTINATION LOYALTY

To measure destination loyalty, the ultimate measure is to check the repeat visitation. Although repeat purchase behavior was introduced in the 1940s, but destination loyalty has not received much focus. There have been some studies where repeat travel was used to assess destination loyalty (Oppermann, 1998). It was also suggested that both repeat travel and recommendations received are actions related to loyalty. Destination loyalty has an advantage that it reduces seasonality in low peak seasons.

4.7.1.6 DESTINATION BRAND ASSOCIATIONS

An important part of destination marketing research is to investigate the relationship between brand image and brand identity. It is of utmost importance that strong brand associations are developed for destinations.

4.7.1.7 PERCEIVED VALUE

Perceived value may include both the actual value as well as value for money. Several service encounters may affect this perceived value. These encounters may be as small as a cup of tea or coffee.

4.7.2 ORGANIZATIONAL PERFORMANCE

Organizational Performance deals with the extent to which an organization achieves its objectives, how appropriate those objectives are and how efficiently they have been implemented.

Some authors like Akehurst et al. (1993) gave certain indicators of organizational performance as follows:

- The number of activities like trade fairs.
- Cost of promotional activities.
- Grants per job created.

Conducting an audit is another method for evaluation of marketing practices. The audit generally examines:

- Changes in the marketing environment at the macro level and operational level
- Assessment of allocation of resources using strategy audit comprising a clear perspective of corporate and marketing objectives.
- Analysis of structure and efficiency of marketing department.
- Analysis of cost-effectiveness.
- Analysis of marketing mix.

At a Conference in 2004, success was summarized by Pacific Asia Travel Association:

- Wrong branding can lead to reduced intention to visit.
- "Interactive travelers" constituted a good part of the destination market.
- Warm and interactive experience lead to a positive brand image.
- Promotional materials used by the campaigners were consistent with the brand and were of good quality.
- Capturing target market, thus leading to increase in arrival of visitors.

4.8 DESTINATION EFFECTIVENESS

The ultimate goal of tourism organizations is to provide such experiences to tourists that they are not able to forget them. This requires consistent and systematic efforts from DMOs to formulate such strategies that lead to this goal. E-marketing, strategic marketing and value chain collectively contribute to the effectiveness of destination marketing. For this, a framework was suggested by Soteriades (2012) where the structure of relationships was clear such that all parties worked with common interest. The effectiveness of these relationships was also dependent on the coordination between the private sector and public sector.

The public sector provides the resources whereas the private sector provides the marketing facilities. It is suggested that use of *e-marketing* can provide added advantage as it is possible to reach people outside one's region with the help of ICTs.

In the context of tourism, in order to provide efficient services, there is a need to coordinate the resources and capabilities of various partnering businesses through cooperation. These businesses cooperate when at a distance from target customers but when they are in proximity, they tend to compete with each other. So, when both *cooperation* and *competition* coexist, it leads to another term "coopetition." In competition, the focus is on value providing strategies whereas in the case of cooperation, the aim is to provide collective strategies to generate value (Wang and Krakover, 2008) (Figure 4.7).

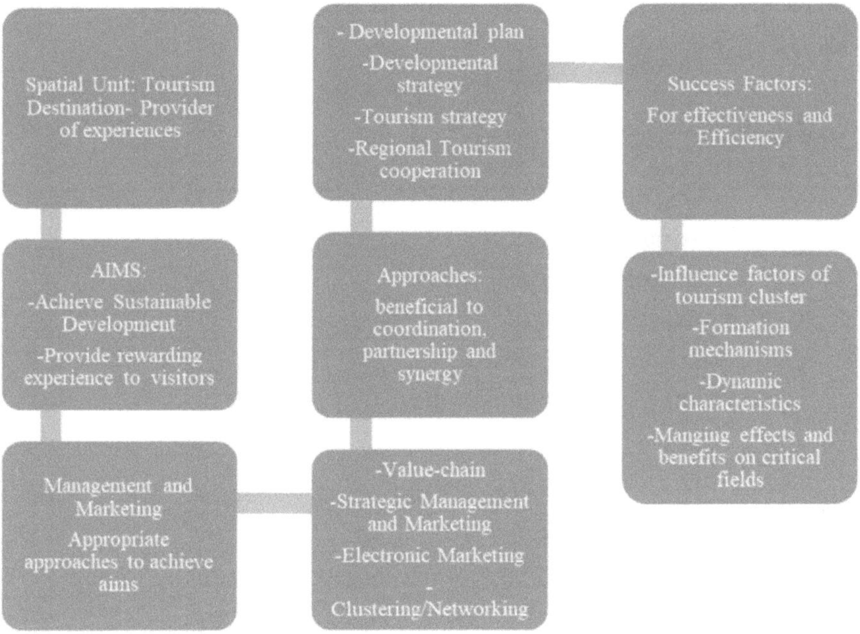

FIGURE 4.7 Destination effectiveness model.
Source: Authors.

4.9 SUMMING UP

A DMO is a form of alliance that involves the coming together of stakeholders within an area or region, for the purpose of promoting and marketing the destinations image and attractions to potential visitors. There are mainly five key stakeholders who include tourists, competitors, community, local tourism, and intermediaries. The key roles of DMOs include marketing, coordination, monitoring quality, and enhancement of relations. Two common approaches used by DMOs to enter a market include leveraging physical and financial assets, leveraging knowledge and skills. The various steps in destination marketing planning include a situation analysis, incorporating a resource audit, environmental analysis and competitive positioning analysis, development of goals and strategy, an action plan and measures of performance. The two broad categories of performance evaluation for DMOs include market performance and organization performance. Lastly, the framework was suggested by Soteriades (2012) where the structure of relationships was clear such that all parties worked with common interest.

KEYWORDS

- **marketing planning**
- **market performance**
- **marcom evaluations**
- **destination effectiveness**
- **consumer-based brand equity**

CHAPTER 5

Destination Marketing Experiences

5.1 DESTINATION EXPERIENTIAL MARKETING

Keller (2013), in his book on Strategic Brand Management, stated that experiential marketing promotes a product by not only communicating a product's features and benefits but also connecting it with unique and interesting consumer experiences. There are five different types of experiences provided to customers which include:

- sense
- feel
- think
- act
- relate

Further, Rathee and Rajain (2019) focused on the importance of experiential marketing in the leisure industry including tourism. They have mentioned extraordinary experiences which lead to experiential value gained at experience touch points. The authors have mentioned that experiential value can be created by including a fun element which makes the experience enjoyable. Earlier, another researcher Volo (2010) had said that **experience marketing** in the context of tourism industry is "the process of promoting, developing, and delivering value experiences to tourists." In tourism industry, experience marketing acts as a link between experiences provided to customers and tourism organizations. For any business to succeed, it is important that they focus on providing memorable experiences. As on the supply side, DMO focus on providing memorable experiences, similarly, on the demand side, customers focus on getting unique and appealing experiences. As such, a tourist's main aim is to get once in a lifetime experience when traveling to a destination.

Destination Marketing: Creating Memorable Tourism Experiences. Rupa Rathee & Pallavi Rajain (Authors)

Since the impact of the concept of **experience economy** can be widely seen in the case of destination marketing, destinations are not viewed as places to visit, but instead viewed as an emotional experience sold for the psychological benefits produced by them. As suggested by Maslow's hierarchy of need, self-actualization lies at the top of the pyramid and gaining meaningful experiences provides scope for personal growth. Even when the experiences gained by tourists are intangible, the means of service delivery are still tangible and thus the consumer is able to fully acknowledge its presence. Thus, tourism marketing can be divided into two separate layers that function together which include the tangible services and the intangible experiences.

As explained by Pine and Gilmore (1999) that the commoditization of goods and services has led experiences of customers to be the priority of organizations. In the case of modern economy, the focus was on features of goods and services whereas in experience economy, the focus has shifted to interactions and emotions. In this context, Holbrook (2006) has stated "consumption is being used to trigger the imagination and realize dreams, to experience intense emotions and engagements and to entertain oneself." The type of experience gained adds value to customer's purchase as the emotions get influenced which effect memory.

Experience marketing rose with the rise in experience economy, and with this change, the promotion and selling of products has also shifted to promotion of experiences gained during consumption. The distinctive features of experience marketing include:

- Focus on customer experience.
- Thinking of consumption as a hedonic experience.
- Recognizing the emotional aspects of products.
- Use of marketing strategies to appeal to the emotional side of the consumers.

Thus, experience economy has a huge impact on the behavior and decision-making of the tourism consumers. Still there has been less focus on studies related to experience in the field of marketing of destinations. Lugosi and Walls (2013) have also suggested that even though experience economy is of utmost importance to marketers, but there is even greater need to create engagement with experiential marketing.

5.2 DESTINATION BRAND EXPERIENCE

Destination brand experience can be explained with the help of the model in Figure 5.1 wherein there are four subsections of the destination brand experience, namely,

- sensory
- affective
- behavioral
- intellectual

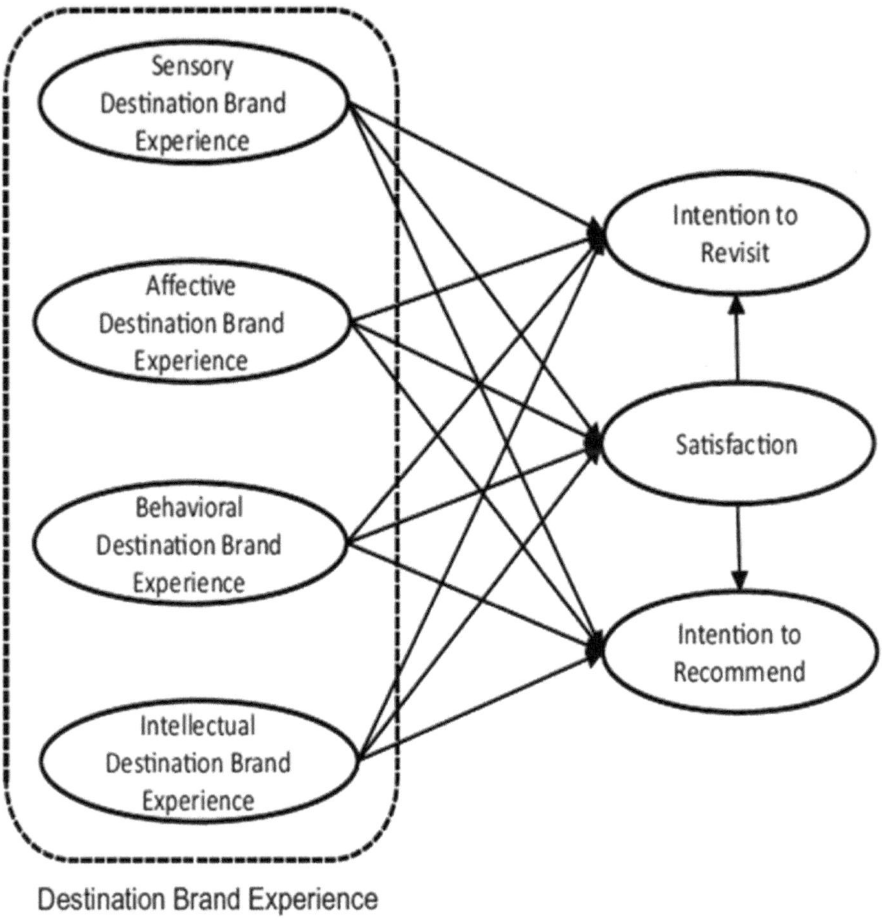

FIGURE 5.1 Destination brand experience.

Source: Reprinted with permission from Barnes et al., 2014. © Elsevier.

Not all these components have an equivalent effect at every destination, as one aspect which may work very well for one destination may not be successful for the others. All the four components have significant influence on intention to recommend and revisit with satisfaction having a mediating role. But, in general, the component that was found to be most effective was sensory destination brand experience. This suggests that senses have the maximum influence on visitor's behavior in the form of tactile, gustatory, aural, olfactory, and visual experiences. For example, when a tourist visits a destination, he:

- Goes sightseeing (visual).
- Enjoys spas and thermal baths (tactile).
- Tastes different types of indigenous food (gustatory).
- Visits various music festivals (aural).
- Visits parks and gardens (olfactory).

Thus, gaining a wholesome sensory experience.

As with any other organization, organizations dealing with destination marketing also face similar challenges of competition from other destinations and even from other sources providing leisure experiences.

Destination experiences are composed of several tangible and intangible purchases. However, intangibles are not purchased conventionally rather they are experienced. This interaction between the tangible destination and tourists along with intangible experiences between tourists and local people form the center of tourism experiences.

Marketers do not have total control over the experience gained by visitors but they influence a limited part of the environment. These experiences occur within the mind of the tourist and the result of these experiences can be seen through the reactions of tourists in various situations. Thus, tourists act as active participants in creation of their own experiences while interacting with

- marketers
- local populations
- sellers
- employees

of various organizations involved. The interaction with local population is quite engaging for tourists and marketing organizations has very less involvement in influencing these interactions. Tourists find local cultural activities joyful and fun leading to additional value to the experiential process.

5.2.1 LIVE-LIKE-A-LOCAL EXPERIENCE

The younger generation likes an experience of living like a local by tasting and experiencing the local culture. There is even an organization by the name of "like a local" that offers such experiences. These experiences include dancing, painting, and art workshops conducted by local people. Another example is couch surfing which provides cultural exchange by arranging stay with locals in place of hotels (Figure 5.2).

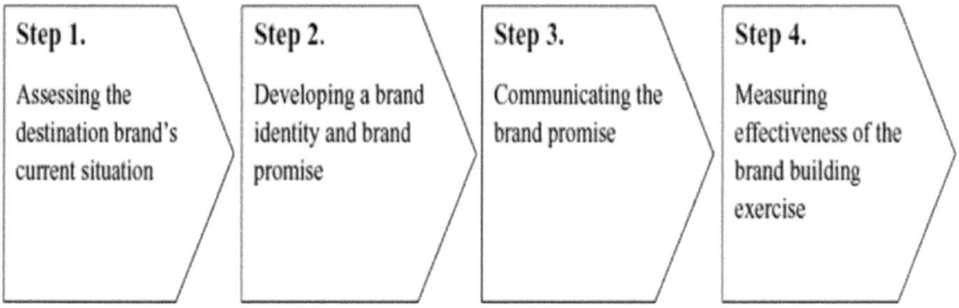

FIGURE 5.2 Steps in building brand experience.

Source: Reprinted with permission from Hudson and Ritchie (2009). © John Wiley and Sons.

5.3 BUILDING BRAND EXPERIENCE FOR A DESTINATION

There are four steps in building the brand experience for a destination:

Firstly, assessment of the situation of particular destination brand.
Second, development of a brand identity.
Next, Communication of this identity to the tourists.
Lastly, measurement of the effectiveness of this process.

Assessment of destination brand situation: For analyzing the situation, it is required that all the key stakeholders which include tourists, DMOs, and tour operators have an understanding of the position of the brand when compared with its competitors. For example, Las Vegas which was previously projected as "Sin City" sent confusing messages when some of the brands portrayed themselves as family brands (Figure 5.3).

Therefore, they took on a new campaign "What happens in here stays here" and this brand planning took around 3 years to complete.

FIGURE 5.3 Re-branding Las Vegas from "Sin City" to "What happens here stays here".
Source: Stefan Wagener - Las Vegas. https://creativecommons.org/licenses/by/2.0/

Development of brand identity: This stage requires that the brand personality should match with the target audience. The brand has both benefits and features suited to the customers' needs which instil a feeling of confidence toward the brand. The brand creates a promise to deliver certain experiences and the tourist expects these experiences to be delivered when he/she reaches the destination.

Communication: This step includes communicating the underlying values of the brand to the customers in the form of advertisements, campaigns, logos, and taglines. As discussed earlier, the promotion that is based on experiential value has greater influence on the decision-making of tourists. For example, "Incredible India" campaign launched by the Indian Government for promoting cultural and spiritual enrichment by travel to India was launched online through the website www.incredibleindia.org. This campaign received high recognition as it got millions of hits per month soon after its launch (Figure 5.4).

Measurement of effectiveness: This step is critical because it keeps a check on whether the brand continually keeps evolving according to the needs of the customers so that it can remain appealing to them. One example of measuring effectiveness of this process is by monitoring the arrivals at a destination. This method was effective in case of Incredible

India campaign where an increase in inflow of more than 15% was seen soon after the launch of the campaign.

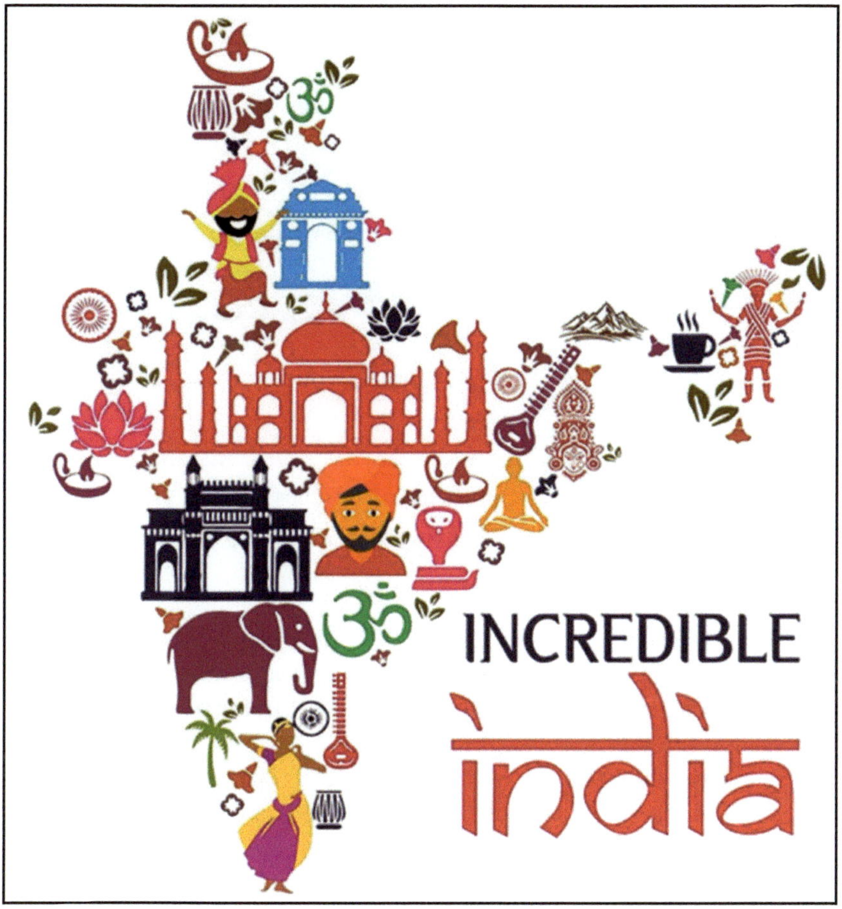

FIGURE 5.4 Incredible India.

5.4 EXPERIENCE CO-CREATION FOR DESTINATIONS

The task of facilitating a space for experience *co-creation* lies with the DMOs, the place needs to be attractive and should allow the creation of valuable experiences. The focus of all activities should be the consumer who helps the suppliers in value creation for memorable experiences. This value co-creation with the involvement of customers in this case tourists leads to growth and innovation.

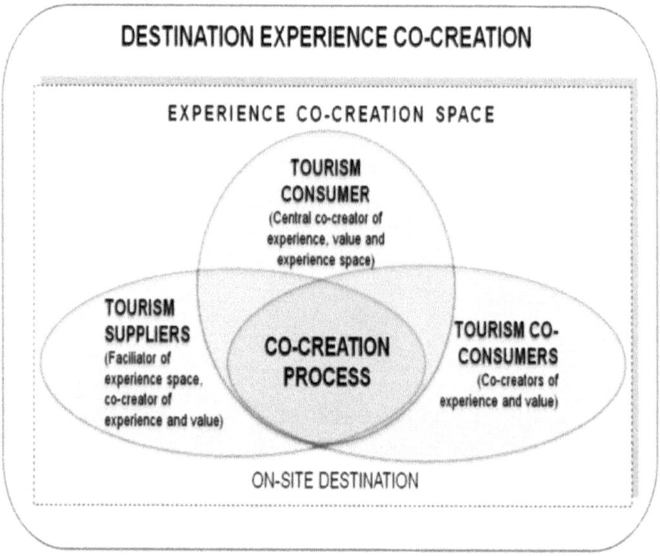

FIGURE 5.5 Model by Prahalad and Ramaswamy (2004).

Source: Adapted from Prahalad and Ramaswamy. © 2004 Wiley Periodicals, Inc. and Direct Marketing Educational Foundation, Inc.

Thus, destinations which follow this approach of value co-creation gain competitive advantage. But the simple model in Figure 5.5 showing value co-creation provided by Prahalad and Ramamswamy (2004) is not so relevant in today's world of technology.

Therefore, the updated model in Figure 5.6 provided by Neuhofer et al. (2012) is more apt. With the rise in ICTs, there is a need to study the impact of technology on experience or in other words, how technology enhances destination experiences.

For example, the popular mobile payment platform in China, Alipay, was used by Visit Finland for its inbound Chinese travelers leading to a widespread adoption. The country has managed to increase the inbound interest in travel to Finland from China by taking the initiative to bring a technology preferred by a growing inbound demographic (Vivion, 2018). South African Tourism, the country's national tourism authority, collaborated with Google's Street View Camera Loan Program in partnership with Drive South Africa, to coordinate the crowdsourced project Google Trekker South Africa. Next, Global pocket Wifi provider, GlocalMe by Hong Kong uCloudlink Network Technology Ltd. enables international travelers to access mobile internet globally with a GlocalMe empowered mobile device at local-comparable or more attractive rates. Similarly, Kyiv City Guide, a mobile application that guides tourists through the most famous locations of Kyiv. The application enables the guests to create their own route, with it even working off-line, when where is no Internet connection (WTM team, 2018).

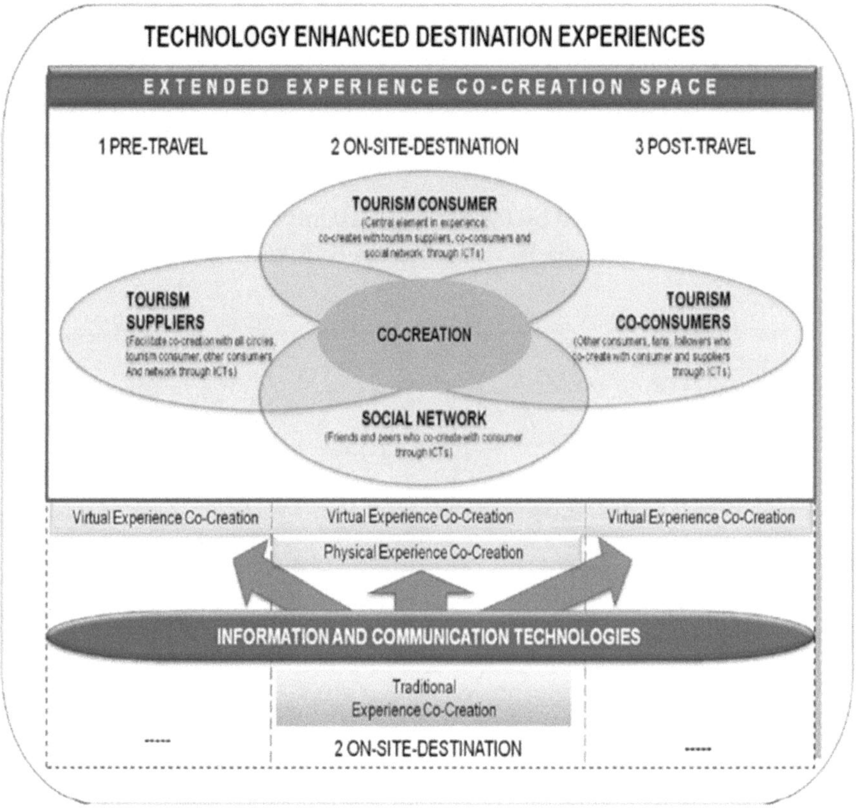

FIGURE 5.6 Experience co-creation.

Source: Borys Kozielski

5.5 MEMORABLE EXPERIENCES

There are seven components of memorable experiences according to Kim et al. (2012) which provide positive experience to tourists. These dimensions include hedonic, novel, meaningful, knowledge, refresh, culture, and involvement.

1. Hedonism deals with the emotional aspect of tourism industry.
2. Novel means some newness or innovation in presenting goods and services to tourists.
3. Meaningful means the satisfaction and fulfilment derived from destination experiences.
4. Knowledge relates to the intellectual development due to the tourists' experiences.
5. Refreshing means providing a feeling of freshness and an environment that is relaxing.
6. Culture means the level of engagement with the local people.
7. Involvement means that something is of personal interest (Kim and Ritchie, 2014).

In the model provided by Packer and Ballantyne (2016), the experiences described are divided into 10 categories as follows (Figure 5.7):

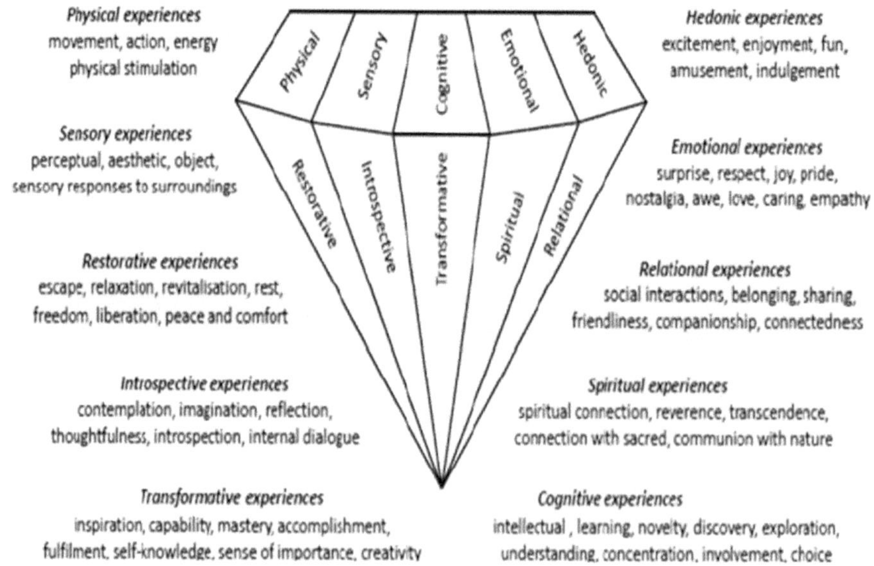

FIGURE 5.7 Visitor experience model.
Source: Reprinted with permission from Packer and Ballantyne (2016).

Physical experiences: This may include physically experienced tourism like adventure tourism. For example, hiking, water rafting, trekking, camping, backpacking, rock climbing, and sky diving (Figure 5.8).

FIGURE 5.8 Physical experiences.

Source: Yogatherapypoint

Sensory experiences: This includes sensory responses to surroundings, such as seeing, listening, smelling, and touching. Example when visiting a destination, a person sees the location, hears music or nature, smells flowers or food, and touches various items (Figure 5.9).

Restorative experiences: This includes experiences related to relaxation, rest, and revitalization. For example, visit to a spa (Figure 5.10).

Introspective experiences: These experiences deal with the inner-self, introspection, and internal dialogue (Figure 5.11).

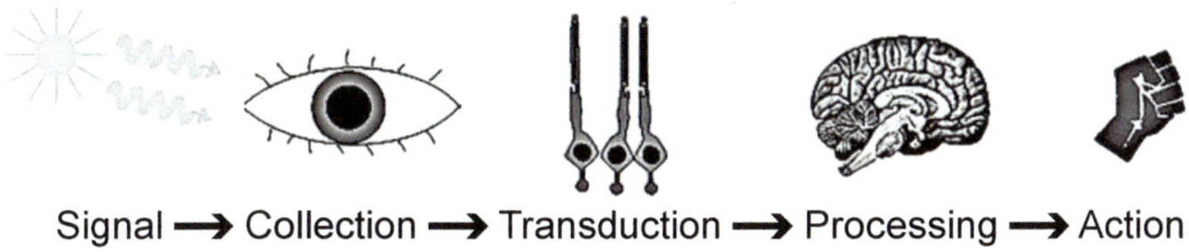

FIGURE 5.9 Sensory experiences.

Source: Thomas.haslwanter - Own work

FIGURE 5.10 Restorative experiences.

Transformative experiences: Such experiences lead to accomplishment, mastery, and creativity for example, learning something new like drawing or painting. Other example like doing something for the first time (horse riding) (Figures 5.12 and 5.13).

FIGURE 5.11 Restorative experiences.
Source: scmp.com

FIGURE 5.12 Transforming experience.

Source: https://pxhere.com/en/photo/1584839?utm_content=shareClip&utm_medium=referral&utm_source=pxhere

FIGURE 5.13 Transformative experiences.

Source: thesprucepets.com

Hedonic Experiences: These experiences require challenge, interactivity, or novelty along with providing fun, enjoyment, and excitement. These experiences are characterized by pleasant or unpleasant sensations. The examples are surfing, paragliding, and skiing (Figure 5.14).

FIGURE 5.14 Hedonic experiences.

Source: Photo by Shalom Jacobovitz. https://creativecommons.org/licenses/by-sa/3.0/deed.en

Relational Experiences: It includes social interactions, connectedness, and companionship (Figure 5.15).

FIGURE 5.15 Relational experiences.

Source: thecentersf.com

Spiritual Experiences: These types of experiences deal with spiritual connection, reverence, and sacred connection (Figure 5.16).

FIGURE 5.16 Spiritual experiences.
Source: Watashiwa.vikram - Own work

Emotional experiences: These experiences relate to a person's emotions leading to awe, nostalgia, and surprise. Trip to Disneyland for kids is an example (Figure 5.17).

FIGURE 5.17 Emotional experiences.
Source: Jtneill - Own work

Cognitive experiences: Experiences related to discovery, exploration, and involvement (Figure 5.18).

These 10 facets may be exhaustive or comprehensive, but still there may be dimensions left unexplored which may need further research. This multifaceted model can be applied to several types of leisure activities and tourism. Measurement activities based on such a model can help to compare different tourist destinations based on positive experiences gained through them. Capturing the essence of visitor experience through each of these dimensions can help to evaluate experiences, measure changes, compare experiences, identify most important factors, engage tourists, explore tourist perceptions, and capitalize on experiences.

FIGURE 5.18 Cognitive experiences.
Source: https://commons.wikimedia.org/wiki/user:Amrita_1830395

5.6 SUMMING UP

Experiential marketing promotes a product by not only communicating a product's features and benefits but also connecting it with unique and interesting consumer experiences. Marketers do not have total control over the experience gained by visitors but they influence a limited part of the environment. There are four subsections of the destination brand experience, namely, sensory, affective, behavioral and intellectual. Further, there are four steps in building the brand experience for a destination: firstly, assessment of the situation of particular destination brand. Next, development of a brand identity and then communication of this identity to the tourists. Lastly, measurement of the effectiveness of this process. There are seven components of memorable experiences which provide positive experience to tourists. These dimensions include hedonic, novel, meaningful, knowledge, refresh, culture, and involvement. In the model provided by Packer and Ballantyne (2016), the experiences described are divided into 10 categories, namely, physical, sensory, restorative, introspective, transformative, hedonic, relational, spiritual, emotional, and cognitive.

KEYWORDS

- **destination marketing experiences**
- **experience economy**
- **experience marketing**
- **destination**
- **brand experience**

CHAPTER 6

Role of Social Media

6.1 DESTINATION MARKETING THROUGH SOCIAL MEDIA

Social media which has the task of connecting people has been used for several other purposes as well like marketing. However, as pointed out by Leung et al. (2013), destination marketers have been slow in adopting social media for promotion of destinations. Hamill et al. (2012) had also found that DMOs in European countries were not fully utilizing the potential of social media for marketing of destinations. While most of them had a Facebook page or a Twitter account, very few of them were able to successfully engage customers through them. The engagement level was measured using influence, intimacy, interaction, and involvement. These researchers also found that DMOs do not use the vast amount of information shared by travelers to the best of their use. If they try to monitor and analyze such content, they may gain strategic knowledge.

Social networking site (SNS) is a platform on the internet which helps to increase social connections between people. It has developed into an important channel for marketing communications. Many researchers have found that SNS is a good option for promoting goods and services through online campaigns (Li, 2007; Zhu, 2013).

6.2 DESTINATION IMAGE AND SOCIAL MEDIA

In this digital era, many aspects of social interactions have changed dramatically. Whether people are looking for information or not, they are being exposed to abundance of information. Earlier information regarding destinations reached tourists through marketers but nowadays everybody

Destination Marketing: Creating Memorable Tourism Experiences. Rupa Rathee & Pallavi Rajain (Authors)

is able to share information about a destination through social media by posting their comments and reviews. Thus, the model given in Figure 6.1 displays the role played by social media is creating the image for a destination.

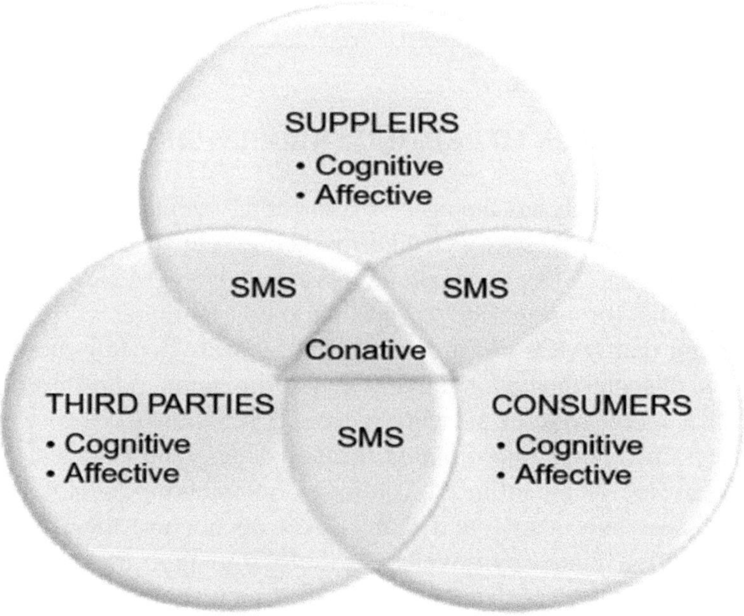

FIGURE 6.1 Destination image formation through social media.
Source: Ghazali and Cai (2014).

Previously, Gartner (1993) had given a framework which focused on the supply side but it was later improved by Tasci and Gartner (2007) which focused on the demand side as well. This model is based on these two previous models. The model in Figure 6.2 develops on the concept that there are three parts of image formation (cognitive, affective, and conative) which are influenced by social media sites.

The model shows that social media is used to connect one party with another, that is,

- Suppliers with consumers.
- Consumers with third parties.
- Third parties with suppliers.

Sometimes destination marketers have complete control over information that they provide to tourists, but such information is uncontrollable due to the involvement of technology. Thus, this model suggests exclusivity is impossible between the three parties in the presence of social media sites.

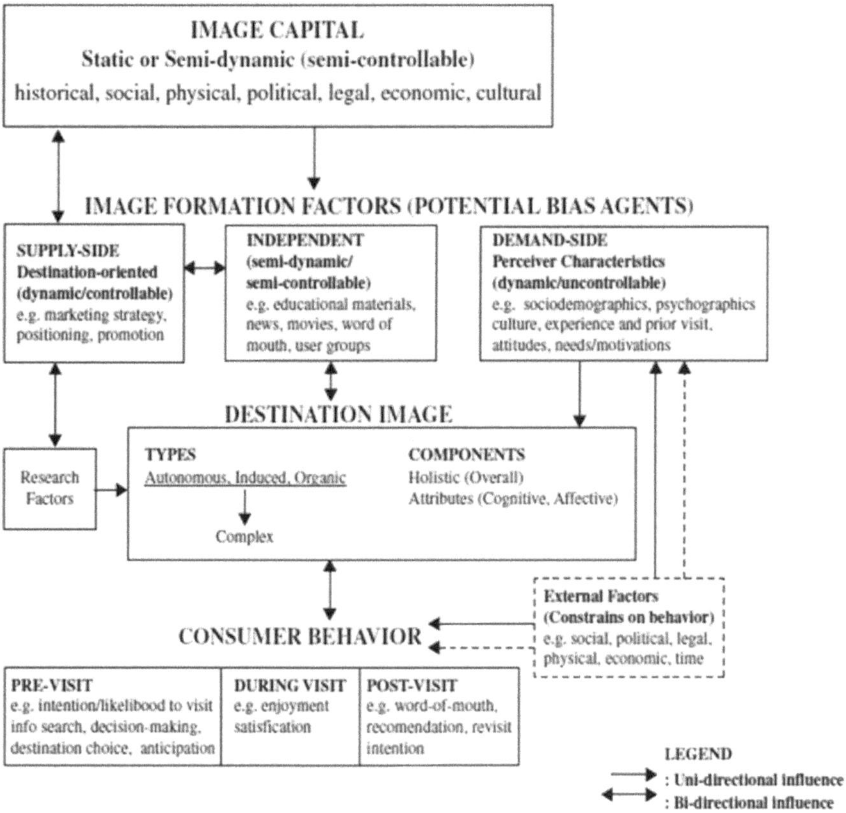

FIGURE 6.2 Destination image.
Source: Reprinted with permission from Tasci and Gartner (2007).

Another aspect that is kept in mind is how the image is formed. There are two basic categorizations of image formation which include organic and induced image formation.

Organic images are formed when tourists form impressions of destinations without actually visiting them. Such images are formed indirectly through sources that do not fall in the category of advertising. On the other hand, **induced images** are promoted by destination marketers through several forms of promotional activities. Even though these two types of images are formed differently, but still they are dependent on each other.

The role of social media in creating images for destination is constantly on the rise. Travelers gather information regarding destinations from social media sites and also form groups to share information.

Several social events and festivals are promoted through social media to develop interest for a destination. This promotion through festivals forms a part of the branding strategies as it provides competitive advantage to destinations in a highly competitive world. Festivals have an influence on the image of the destination being promoted through social media (Garay and Perez, 2017).

6.3 IMPORTANCE OF SOCIAL MEDIA

Most of the social media have groups or communities where people can share their similar interests and thoughts. Tourism is one of the most popular topics discussed among groups on social media. The ease with which information can be shared on social media has led to its growth. Some of common social media like SNS, blogging sites, and photo sharing sites are among those that provide space for sharing information and recommendations. Social media acts as a platform for searching information and planning the trips. The **viral marketing** done through Facebook and Twitter is an example of the important role played by social media.

The way people used to plan their trips has changed with the advent of social media. It was found that more than 80% of travelers planned their trips using the internet (Guitart et al., 2010). Social traveling sites like TripAdvisor, Make My Trip, Yatra.com, Cleartrip are among the ones who allow the tourists to experience what other tourists have experienced in the phase before traveling and this is done through their websites. The *reviews* provided by other travelers are helpful in making travel decisions for a destination. The users of these sites can post *comments* and *ratings* which help other users regarding the choice of hotel, attractions, and other tourist services (Figures 6.3–6.19).

The social media sites have the capability to reach potential customers, in this case, tourists by working effectively as a means of communication and advertising. It connects people with similar interests and they have a choice to even come together and plan a trip. Social media is beneficial in propagating information about a destination while eventually helping the whole country economically by increase in tourism.

FIGURE 6.3 Tripadvisor logo.
Source: Trip Advisor

FIGURE 6.4 Tripadvisor review.
Source: tripadvisor.com

FIGURE 6.5 Indian travel company.
Source: Sterling Holidays Resorts Limited.

The most popular social networking sites for destination marketing include Facebook, Twitter, Instagram, Blogs, and Official websites.

6.3.1 FACEBOOK

Facebook is a social networking service that was founded in United States of America in 2004. It has 2.90 billion users as of January 2022. In India, there were 349.2 million Facebook users as of January 2022 (Statista.com, January 2022). The service has all kinds of users all over the world who connect with other people by sending Friend Requests. Some organizations have Facebook pages and by joining these pages, users can receive updates on whatever new is happening. Among various SNSs, Facebook is the most popular and a vast majority of its users get inspiration for travel and holidays through interaction with tourism organizations. Researchers have suggested that the outcomes of Facebook's marketing efforts can be understood by measuring the impact that the Facebook pages have on the behavior of customers (Tuten et al., 2014).

In their study on the impact of Facebook pages of destinations on tourist's attitude and intention, Leung and Jiang (2018) found that social influences had a significant impact on intention to visit. Such findings are quite useful for DMOs as they can put marketing efforts through Facebook pages to promote a destination.

Use of Facebook by DMOs:

- DMOs highlight their value systems on their Facebook pages. If these values are in coherence with the values of the tourists, they would be attracted. For example, DMOs depicting sustainability through their pages can attract tourists who have more environment friendly and sustainable goals.
- The detailed information about destination that is available on Facebook pages also helps to remove uncertainty and anxiety from tourist's mind.
- The advertisements available on Facebook pages regarding various destinations when found appealing and engaging tend to attract many travelers. As suggested by the theory of planned behavior, attitudes, subjective norms, and planned behavior control have influence on the behavioral intentions of customers.
- Another factor that influences the future intentions is the past experience.

FIGURE 6.6 Marketing of destination on Facebook.

6.3.2 TWITTER

Twitter is a social networking site that is used for posting comments in the form of "tweets." It was started in 2006 in United States of America and has an estimated 396 million users as of January 2022. In India, there were 22 million twitter users as of January 2022 (statista.com, January 2022). Twitter uses "tweets" or short messages up to 140 characters which was doubled to 280 characters in 2017. The registered users can post-tweets and re-tweets, but others can only read these tweets. Twitter has a wide variety of users ranging from individuals to organizations, celebrities, politicians, etc.

Use of Twitter by DMOs:

- DMOs use Twitter for promotion of destinations. Official tourism organizations that have a large number of followers also use Twitter as a means of promoting destinations and increasing tourism. Through Twitter, active users can connect with others by either including links in their tweets, re-tweeting or using "hashtags" which are popular these days.

- Twitter has the ability as a social networking site to enhance the image of a destination by sharing information about locally held events, mentioning opportunities for DMOs and broadcasting messages to a large group of people.
- Twitter can be used by DMOs to create awareness about their brand, promotion of images and build relationships. Twitter is an effective means to promote destinations through electronic word of mouth (e-WOM). As most of tourist's choice of destination are influenced by word of mouth, so Twitter rightly serves this purpose by marketing through e-WOM.

Antoniadis et al. (2015) found that there was a direct relationship between tourism performance and Twitter performance. They found that accounts with high Twitter performance indexes showed high tourism performance indexes.

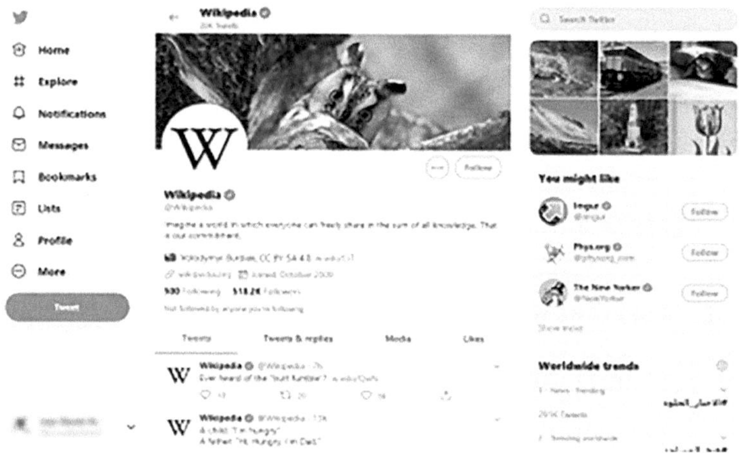

FIGURE 6.7 Destination marketing on Twitter.

Source: Twitter (& Wikimedia Foundation) - Screenshot of [2], taken 25 May 2020 around 9 pm PST

Use of Twitter is an inexpensive means of communication for DMOs. They have been using Twitter for the past decade to share information about destinations. Since the number of users on twitter has increased tremendously over the years, it is necessary to grasp the attention of users. DMOs face tough competition from other marketers who are trying to seek the attention of users. One way to garner the attention of users is to make use of influential users. The level of influential users can be measured by number of followers, re-tweets, and some other characteristics.

6.3.3 INSTAGRAM

It is a popular social networking service for sharing videos and photos, currently owned by Facebook. It was originally launched in 2010 and was acquired by Facebook in 2012 US$ 1 billion. After its launch in 2010, Instagram gained popularity rapidly with 1.3 billion registered users as of January 2022. The application has several features like geographical tagging, editing uploaded media using filters, and organizing data by using *hashtags*. The posts on Instagram can be shared publicly or with selected *followers*. Users can browse through other user's content as they look through the *trending* content. There is option of liking other users' photographs and adding content to a feed. As of July 2020, Christiano Ronaldo (the footballer) has the largest number of followers.

Instagram is being used widely for advertising since November 2013 when the first image ad appeared in its feed and almost a year later in October 2014 the video ads appeared. Instagram has more than 1 million advertisers on its platform. In 2019, Instagram allowed the feature of selling products directly through its app by adding "Checkout on Instagram" option.

6.3.4 BLOGS

It is another powerful tool for the development of relationships with potential tourists. Blogs as defined by Sharda and Ponnada (2008) are "virtual diaries that are stored on the web and are available to anyone." They may be in the form of audio, video, pictures, or text. Blogs or in the case of destination marketing "travel blogs" has become a popular source of travel information. These blogs can be shared among family, friends, and so on to turn them into potential visitors or in near future as new visitors to destinations being shared.

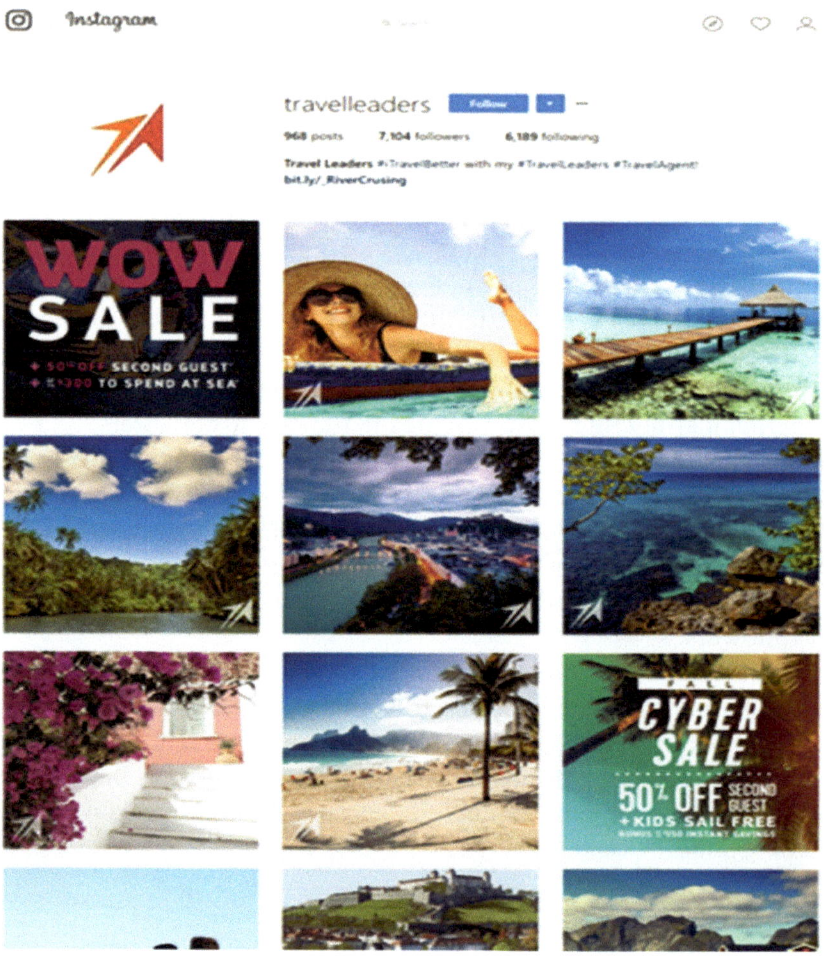

FIGURE 6.8 Destination marketing on Instagram.

Use of Blogs by DMOs:

- These travel blogs are also useful for DMOs as they can monitor the attitude of visitors regarding a particular destination.
- They help DMOs in modifying their strategies to suit the needs of the visitors according to their profiles. A successful example of blog campaign was the destination marketing of South Africa. They collaborated with 15 international bloggers for the promotion of their destinations. It has been promoted as "#meetsouthafrica"

campaign circulated widely through travel blogs (Ontiveros, 2016).

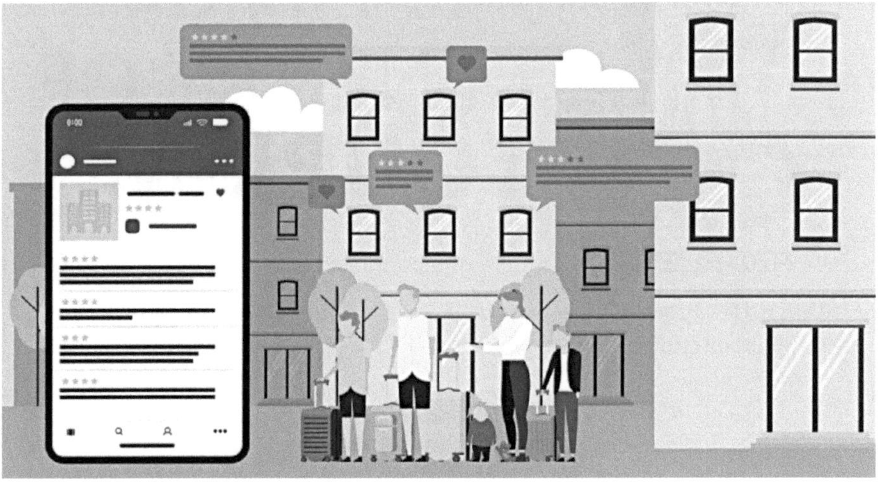

FIGURE 6.9 Travel blog.
Source: Familydestinationsguide.com

A different marketing approach was used by Kerala Tourism with a campaign by the Kerala Tourism department known as "Kerala Blog Express." The event was a huge success and helped in reaching out to aspiring travelers who wanted to travel to Kerala (Choudhary and Dixit, 2019; KeralaBloggers are all in praise of Kerala, 2018).

6.3.5 OFFICIAL WEBSITES

The government or tourism departments generally make use of official websites to promote tourism. These websites include several types of information, such as travel, accommodation, and sightseeing. The websites generally use the terms such as visit, go, travel, enjoy, for example, visit-greenland.com, ecuador.travel, keepexploring.canada.travel, etc. These websites make use of captivating videos, visual experiences that are immersive, page elements that are dynamic and can be integrated with social media.

पर्यटन मंत्रालय
MINISTRY OF
TOURISM
सत्यमेव जयते

FIGURE 6.10 Indian tourism website.
Source: Swapnil1101

FIGURE 6.11 Incredible India.
Source: Incredibleindia.org

6.4 INFLUENCER IDENTIFICATION

In their research, Bokunewicz and Shulman (2017) tried to find the degree of influence by different accounts on Twitter and how that can be identified. The researchers suggested that every social network has key players or influential people in their networks. Identifying important or influential people is key to have an impactful presence on the network.

This influence can be assessed through four metrices:

- First, in-degree, it can be assessed by the amount of attention received by account holders. In the case of twitter, this is assessed by finding the number of people who replied or mentioned a particular account. Thus, in simple terms it can be referred to as the popularity of an account.
- Secondly, out-degree, which is measured by the number of accounts to which a particular account broadcasts or in the case of Twitter, the number of arrows pointing away.
- Thirdly, eigenvector centrality, which overcomes the limitations of degree metrices. In the case of degree, each connection holds equal importance whether it is influential connection or an unknown person. But with eigenvector centrality, the advantage gained with inclusion of influential connections can also be added.
- The last metric is the betweenness centrality. This can be understood by the fact that on social networks, information moves through contacts, that is, from one contact to another and so on. For example, on Twitter, when information moves through many short paths, it suggests large betweenness centrality.

6.4.1 FOCUS AREAS FOR SOCIAL MEDIA

The core areas that the social media focuses on for marketing of destinations include food, entertainment, shopping and travel.

- In a study by Shen et al. (2015), diet was found to be the most important factor as the Singaporean special food was prime for the image formation of this destination. The different cuisines form an integral part of Singapore's culture.
- Second comes the travel or tourist attractions.
- Shopping is another important factor that attracts tourists to various destinations all over the world. SNSs also primarily focus on shopping sites at a particular destination and provide a range of products from low end to high end in order to cater to all kinds of travelers.
- Lastly, entertainment or leisure activities also contribute to a large extent in order to promote a destination. The famous personalities of that region also help in promotion through advertisements and live events.

Besides these four focus areas, some other factors include culture, fashion, transportation, and sightseeing. The official accounts of DMOs on SNS mainly focus on scenic spots and food of the destination besides a positive environment and friendly culture of that region.

6.5 DESIRE AND INTENTION TO VISIT

The belief-desire-intention (BDI) model by Bratman et al. (1988) explained the behavioral intentions of people. In this model, desire is a determinant of intention as intentions are caused by behavioral motivations. In simple terms, desire depicts the outcomes that would be achieved through the behavior. For example, people have beliefs formed on the basis of stimuli received through their senses. These beliefs lead to desires which in turn lead to intention to perform a certain action. The BDI (diagram or description) model has been used by several researchers to explain the behavior of consumers. This model can also be used to explain the behavior of destination travelers exposed to different social media. The various media whether mass media or social media affect the attitude as well as behavior of travelers. The vast amount of cultural content to which the travelers are exposed to through social media arouses their feelings and develops a desire in them to visit a destination (Koo et al., 2016).

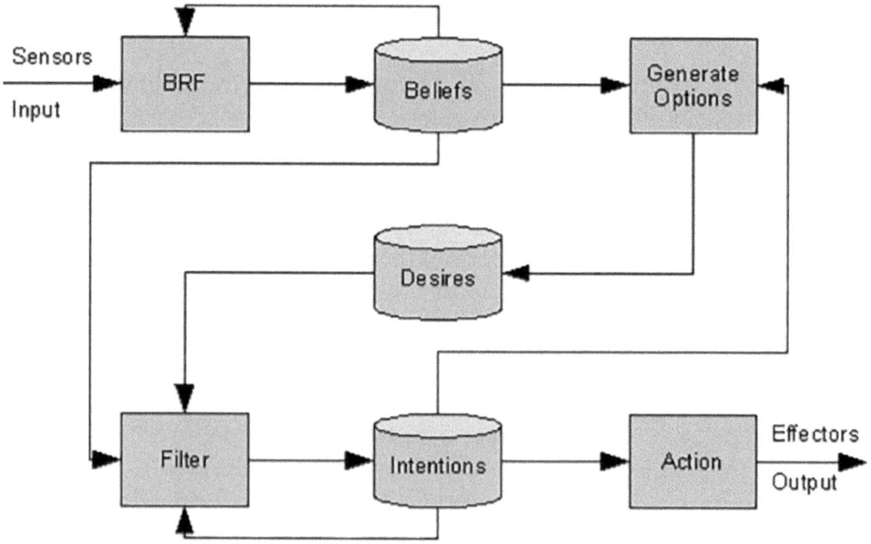

FIGURE 6.12 BDI model.

The rising popularity of social media has made it a good source for travel-related information. The information shared by travelers on these media is used by inexperienced and potential tourists to choose a destination. Thus, the content available on this powerful media is used to attract tourist toward a destination. The BDI model based on previous research suggests that consumer's perceptions based on the exposure to various media develop a desire in them to visit a destination which can also be called as the perceived objective. The constant exposure to such content can lead to greater desires of visiting a destination which suggests that content gratification can lead to inclination to visit.

6.6 INTEGRATING SOCIAL MEDIA FOR DESTINATION MARKETING

A study by Kavoura and Stavrianeas (2015) found that the top three social media used by respondents for searching destinations included official websites, blogs about the destination, and Facebook. The digital age has provided such platforms to people which can have two-way interactions, thus allowing travelers to fully exploit these resources for gathering information. These social media depict different types of uses and relationships between its users. For example, for personal uses, YouTube and Picasa are used, for friendly relations, Facebook and MySpace are used, for creating professional links, LinkedIn has been used. Research suggests that even though social media plays such an important role, it is still not being used to its full potential by marketers with social media accounting for even less than 1%of the marketing budgets.

Gretzel and Yoo (2013) in their study focused on how the four P's of marketing have been used by social media for marketing destinations. In the book edited by Sigala, Christou and Gretzel (2012), the researchers mainly dealt with social media and how Web 2.0 is used by DMO's to market their services either through social media or discussion forums.

Researchers have found that social media helps to create a positive attitude toward a destination and creates an urge in travelers to visit a particular destination. Uses and gratification theory also suggests that content gratification builds a desire in consumer to travel and consequently the consumer tends to revisit a destination (Koo et al., 2016).

Several researchers, for example, Di Pietro et al. (2012) made use of the Technology Adoption Model (TAM) for enhancing the usability, easiness, and fun part of the social networking for promoting tourism. The communication in the form of electronic word-of-mouth is also part of communication strategy that is advantageous over the traditional communication systems. A sense of trust and feeling of community is developed through comments posted online through social media and it has the same effect as direct face-to-face communication.

The widespread success of social media sites has been possible through the users need for communication and socialization. These platforms were easily adopted by various service and manufacturing sector industries to reach the customers.

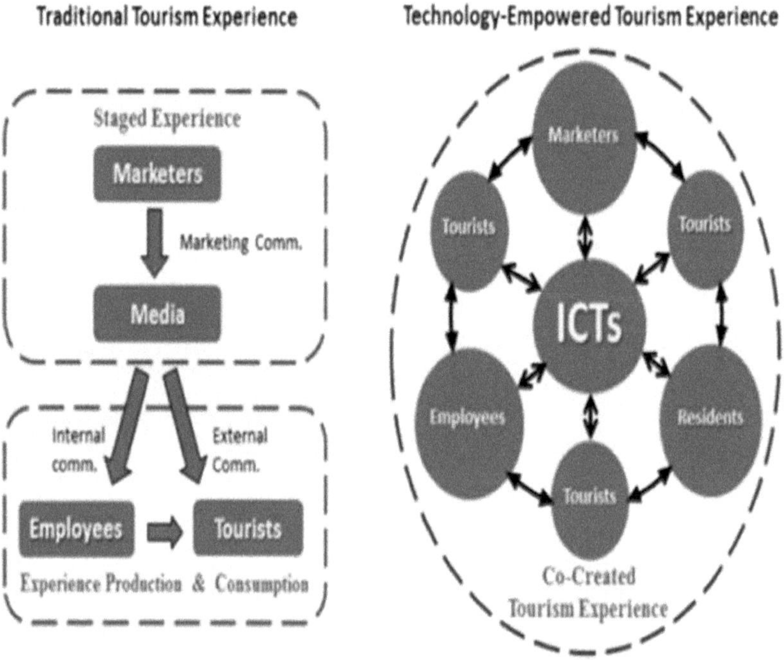

FIGURE 6.13 Difference between traditional and technology-driven tourism.

Source: Reprinted with permission from Yoo and Gretzel (2016). © Emerald Publishing. Limited all rights reserved.

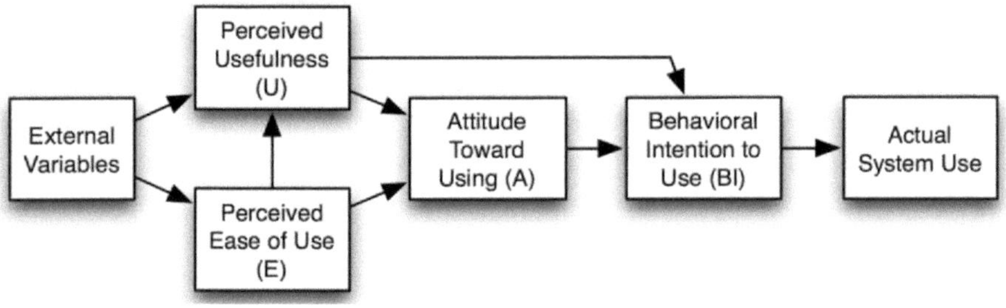

FIGURE 6.14 TAM model.
Source: Adapted from https://commons.wikimedia.org/wiki/File:TAM_v1.svg

6.7 MAKING DESTINATIONS VIRAL

Online resources have developed a huge amount of interest among various researchers regarding various aspects including destination marketing. The online media have transformed the communication to a great level with minimal or no cost being involved in broadcasting messages. The traditional media resources were bound by a cascading flow of information with lot of intermediaries involved, but social media lets the information flow directly to the customer.

Social media have also blurred the lines differentiating between the service providers and customers with them working together in a community as a whole. The online communities work by engaging the customers to a high level as their survival depend on this. Similarly, in the case of destination marketing, place branding cannot be said as the task of destination marketers or managers rather there is a huge involvement of public or customers in branding the place (Sevin, 2013). The people visiting place share information among their networks which if liked is further reshared leading to the information being propagated among various networks thus becoming viral.

Example: A funny video featuring comedian Di Henwood with the campaign centered on the Air New Zealand Facebook page. They have even produced a Facebook application for the spooning campaign, so fans can select some of their friends to spoon and enter a competition to win a weekend away…the marketing message for this campaign was “It’s all about their new Economy Skycouch which allows passengers to lie down during flight to sleep, and presumably spoon if traveling with a close friend.”

Another example was where a clever couple of videos have been produced and placed on YouTube with the star of the show being the likeable Bazza from Oz. Targeted at Australians to visit Northland, these two videos (A language Guide and an Outdoor Guide) were being spread virally through You Tube, Facebook, and Twitter. Bazza has his own Facebook page and Twitter account, so visitors could follow his travels through Northland (Ackers, 2010).

6.8 TECH SAVVY DESTINATION MARKETING

Personalization of digital marketing can be done with the help of digital footprints of tourists. Researchers have suggested that for the prediction of hotel occupancy, the destination websites can be used to gather web traffic data. Differentiating factor based on the ability of knowledge management through technology can lead to competitive advantage for a destination.

However, DMOs lag in adapting to new technology when compared with advanced websites developed by tourism operators. Previous studies have tried to find the challenges faced by DMOs in acceptance of new technologies. Several DMOs have pointed out that they are similar to small organizations on the basis of structure as they too have small staffs which are not technologically savvy. This bottleneck can only be overcome by getting the staff ready for e-business management. This lack of knowledge leads to DMOs using technology only for the purpose of marketing.

The concept of **"smart destinations"** can only be fulfilled when technology is used for understanding patterns in tourist activities with the help of big data analytics. Destination marketing intelligence can be developed by extracting data from user-generated content. The user-generated data has the capability of providing useful information for destination marketing. It will also help to provide personalized information regarding travel to travelers in turn enhancing their experiences.

The researchers focused only on tourism websites whereas there was huge scope for mobile apps for tourists on the go. The exponential growth in the smart phone market has changed the face of tourism organizations as well. Smart phones not only provide information before the travel actually happens but it assists travelers in facilitating their trips by bringing the virtual world in their palms (Li et al., 2017).

6.8.1 VIRTUAL REALITY IN DESTINATION MARKETING

The increasing use of virtual reality in tourism has shown the acceptance it has gained from travelers. This new technology plays a role in the three stages of travel: pre-visit, on-site, and post-visit.

The **pre-visit phase** comprises of development of expectations regarding a destination and decisions related to visiting that destination. In this stage, technology can be used to promote a place in new and innovative ways. The most common ways of doing this is through the destination's website or social media. Augmented reality is another addition to these platforms in promoting destinations. Besides this, virtual reality serves in providing relevant information to the travelers and thus taking informed decisions.

FIGURE 6.15 Use of augmented reality in pre-visit phase for decision-making.
Source: AR Trends

In the **on-site stage,** the user actually experiences the destination through the use of virtual reality. Here, the person puts on the virtual reality device which is either a helmet or goggles to experience the destination while actually being someplace else (explained in detail in subsequent paragraphs).

FIGURE 6.16 Virtual reality tourism.

In the **post-visit phase,** the person after experiencing the destination with the help of virtual reality devices, posts his reviews or experiences in the form of blogs, pictures, or videos.

According to Schroeder (2008), virtual worlds are defined as "persistent virtual environments in which people experience others as being there with them and where they can interact with them." The virtual tools used in virtual worlds provide education and entertainment in an immersive and engaging way. Virtual reality has been particularly appreciated for providing engaging experiences.

Augmented reality and virtual reality can be differentiated as the former provides digital content in real world whereas the latter is psychologically engaging in virtual world. The content is available through head-mounted displays (HMD) in the form of helmets, goggles, for example, those developed by Oculus and Samsung. These devices provide 3D views with such textures and light effect, that it can make any destination more attractive to visit (Marasco et al., 2018).

Previous authors (Chen and Lin, 2012; Huang et al., 2012) have studied the effects of 3D tour itinerary promotion on customer satisfaction, experience, and behavior. The 3D experiences have motivated tourists to visit the destination in real life. The next generation of virtual reality devices has focused the attention of researchers on the use of these devices, their acceptance, and intentions toward them.

6.9 SUMMING UP

Social networking site (SNS) is a platform on the internet which helps to increase social connections between people. Earlier information regarding destinations reached tourists through marketers but nowadays everybody is able to share information about a destination through social media by posting their comments and reviews. The ease with which information can be shared on social media has led to its growth. The most popular social networking sites for destination marketing include Facebook, Twitter, Instagram, Blogs, and Official websites. Every social network has key players or influential people in their networks. Identifying important or influential people is key to have an impactful presence on the network. Social media have also blurred the lines differentiating between the service providers and customers with them working together in a community as a whole. Further, the concept of "smart destinations" can only be fulfilled when technology is used for understanding patterns in tourist activities with the help of big data analytics. Destination marketing intelligence can be developed by extracting data from user-generated content. Personalization of digital marketing can be done with the help of digital footprints of tourists. The increasing use of virtual reality in tourism is also a thing to lookout in the future as it has shown the acceptance gained from travelers.

KEYWORDS

- **social media**
- **destination image**
- **influencer**
- **identification**
- **smart destinations**

CHAPTER 7

Destination Marketing Strategies

7.1 SEGMENTATION, TARGETING, AND POSITIONING STRATEGIES

People do not have same preferences, and companies may find it difficult to satisfy each and every customer. This is true in the case of destinations as well. Thus, there is requirement of proper segmentation, targeting, and positioning strategies to market the destinations in such a way that all customers (tourists) are happy.

7.1.1 SEGMENTATION

The marketers should conduct *extensive quantitative research* for effective segmentation which is based on demographic, geographic, psychographic, and behavioral characteristics.

The destination marketers and travel brands need to make use of every opportunity as *online marketing* has made it easier to reach potential customers. The online marketing suggests that along with "who" to reach, the "when" to reach is also of equal importance. Google has used a term *"micromoment"* while using its search engine which means an urge to immediately answer to user's need. For example, when a user feels compelled or has a desire to visit a destination then he looks for that destination online. If he gets satisfactory information then decision is made, otherwise, he reconsiders his options (Sherwood, 2018).

When making a strategy focusing on a particular segment, *personalization* can be done, which means going a deeper level or aiming at the individuals. For this, travel companies can maintain a database of their users with their

Destination Marketing: Creating Memorable Tourism Experiences. Rupa Rathee & Pallavi Rajain (Authors)

search history so that they can make personalized recommendations. Thus, providing them a feeling of being treated as an individual.

7.1.2 TARGETING

This is another important aspect for developing strategy for marketing destination. The relevant question for developing a targeting strategy is to identify who would like to visit a destination and for what reason. For example, the marketers may target youngsters taking a break from work or elderly people on retirement. There are people who would enjoy sporting activities or other who may just want to relax. There is a need to target each of the different groups which include business travelers, students, or tourists with different interests.

7.1.3 POSITIONING

Positioning requires showcasing the destination with a particular quality or attraction. For example, positioning as a *natural attraction* beautiful gardens, lakes, springs having a pleasant climate. Some destinations have *adventure and sports*, such as mountaineering, skiing, rafting, and golfing. Then there are sites with *cultural activities*, such as festivals, events, and pilgrimages. Shopping and dining included local *art and craft and the cuisine* specific to that place. Positioning the destination as a safe spot is also important especially in the case of destinations, such as Kashmir where acts of terrorism are frequent. *Friendly, peaceful*, and places with good quality of food and hotels are also preferred. Lastly, a *clean environment* with *easy accessibility* brings more tourists (Saqib, 2019).

The marketers need to focus on the *Unique Selling Proposition* (USP) of a destination. The above-mentioned qualities or attractions help in deciding the uniqueness of a location.

7.2 MARKETING MIX STRATEGIES

The marketing mix in the case of services comprises the 7Ps. Each of these Ps plays an important role in providing a memorable experience to the destination visitors. The strategies related to each of these elements are discussed below:

7.2.1 PRODUCT

The destination marketers can describe the product mix with the help of five A's as given by Cooper et al. (1993) which are as follows:

- Attractions (natural and man-made).
- Accessibility (roads, railways, waterways, airways, etc.).
- Accommodation (hotels, guesthouses).
- Amenities (infrastructure).
- Ancillary Services (entertainment).

Generally, tour operators provide all these services together in a package, although Destination Marketing Organizations (DMOs) do not play a role in such kind of bundling services, but they can suggest how the products can be combined.

7.2.2 PRICING

Usually, a new destination is always introduced at a *low price* in order to penetrate the market, with subsequent rise in price as the loyalty increases. A *high-priced policy* is generally used when *skimming* the market or restricting the product to an elite class. The *differential pricing* system is used when there are seasonal variations. Lastly, *competitive pricing* is used in relation to the competitors.

The destination marketers must know the *monetary and the nonmonetary cost* a customer pays. The nonmonetary cost includes the

- Time expenditures
- Physical effort (e.g., fatigue, discomfort).
- Psychological burdens (mental effort, negative feelings).
- Negative sensory burdens (unpleasant sensations affecting any of the five senses).

To increase the net value of the service, the nonmonetary costs of service must be reduced. The marketer can:

- Reduce time costs of service at each stage.
- Minimize unwanted psychological costs of service.
- Eliminate unwanted physical costs of service.
- Decrease unpleasant sensory costs of service.

For effective revenue management *cross-selling* can be done. Cross-selling means the company's other products that are sold to the customer. For example, ski resorts can cross-sell ski lessons and dinner sleigh rides.

Upselling can also be used for effective revenue management. Upselling means training sales and reservation employees to offer continuously high-priced product that will better meet the customer's needs, rather than setting the lowest price. For example, the destination marketers can train front desk employees and incentivize them to upsell guests into suites and premium rooms.

7.2.3 PLACE

Distribution or distribution channels are a part of the option to buy since consumption takes place as the product is being produced. The distribution channels require an understanding of the influence of the trade on the purchase decision of tourists. This is important for destinations in order to exploit the budget of intermediaries. In the digital age, intermediaries have lost their hold as the consumer is connected directly through the Internet. This has changed the role of DMOs from product *owners to facilitators*.

When the market is fragmented with many travel agents, they can develop a *collaborative system* which provides easy access for reservations (Hartl, 2002).

The marketers can reach their audience through *social media* channels by reaching out to people who visited the *destination website*. This allows the marketers to give a reminder to potential customers who had shown an interest on previous occasions. This can be quite helpful in encouraging customers for making a commitment to visit a destination (Revfine.com, n.d.).

Consortia and reservation systems can be developed to promote a specific destination as tourism attractor. A consortium is a group of hospitality organizations that is allied for the mutual benefits of the members.

7.2.4 PROMOTION

Society is transforming. Traditional media like newspapers, magazines, FM radio and TV may not attract the young generation. So, in today's digital era, the destination marketers should focus on smart phones, tablets, *social networks, GPS (Global Positioning Systems) apps, booking engines* to economically reach consumers globally and 24/7.

The destination marketers need to plan thoroughly and allocate the *promotional budget to the new emerging media* like digital and social media rather than only the traditional media.

The advertising should be *innovative and tempting* for the target audience. The visuals of the advertising should be very *appealing* so that it attracts the potential customers to visit the destination. The advertising should also highlight the *special offers or seasonal deals*, for example, hotels in Goa provide deals in rainy seasons. Through *sales promotion*, offers like "Win a tour to Singapore" or "Scratch the card and earn a trip to Paris" can be provided to attract customers.

Direct selling can be done by tour operators through corporate who are selling cruises. They can provide information about special services for example, new timings of airlines during holiday rush period.

New audiences can be reached, for example, by encouraging school children for mountaineering during summer breaks. Another example is of Uttaranchal Tourism luring adventurous tourists to visit Rishikesh for river rafting.

Broadcasting television shows and documentaries showcasing nature and wildlife, for example, National Geographic and Discovery channel show documentary on various destinations.

7.2.5 PEOPLE

The kind of people employed can make or break a business. Investing in the *right kind of people* can be quite beneficial for tourism companies. Hiring *positive and enthusiastic* employees to serve as guides or travel agents brings in more customers. The people employed in the frontline who directly deal with the customers must have a *balance of technical and human skills* so that they help tourists to create a pleasant and memorable experience.

7.2.6 PROCESS

The process of destination marketing includes *blueprinting* the various tasks to be performed while delivering the services. For example, process for a stay at a hotel includes arrival, check-in, spending the time, eating food, check-out. The destination marketers need to ensure that every task in the process in done on time. The service providers need to be in control of the various processes. For this, the *waiting times should be minimized* so that the customers do not have to wait and are delivered the service experience as promised (Ng, 2018).

7.2.7 PHYSICAL EVIDENCE

An important strategy regarding physical evidence which includes building, infrastructure, parking, materials is to *keep them updated.* These should not be outdated as the outdated materials lead to dissatisfaction among customers. Also, the *detailed information* should be available in the brochures of travel companies and tour packages so that the travelers are well informed. The brochures should also be *appealing* as well as well-made so that they are comprehensible to the tourists and available in the *language of their choice.*

7.3 BRANDING STRATEGIES

While building destination branding strategies, the following things need to be kept in mind

Know the perception of the destination

When a tourist visits a destination for the first time, he/she has certain perceptions regarding that destination. For example, when visiting Egypt, people may expect to ride on camels through the desert or when in London, there might be a likelihood of rain. These perceptions may or may not be true. Therefore, it is best to know the perception so that it can be changed or amplified into positive ones.

Capture the essence of the destination

An effective branding strategy is to convey the qualities of the destination to the tourists. It involves peeping into the minds of tourists as they want to escape their everyday life and experience something novel. The success of the branding strategy depends on experience gained at every point of interaction. This interaction begins from the moment the tourists start thinking of visiting the location to the final moment when they arrive. A successful example of this is Tourism Australia which highlights the land, the people, and the wildlife which can be seen nowhere else.

Build on media references for the destination

The media tie-ups serve a powerful strategy for branding destinations. It helps to showcase the culture and beauty of a destination through popular media. For example, VisitBritain which is a tourism organization in UK has used the James Bond movie series to urge the tourists to come to Britain and "live like Bond" (Carter, 2014).

7.3.1 CRITERIA FOR CHOOSING BRAND ELEMENTS

There are broadly six criteria for choosing brand elements which include memorable, meaningful, likable, transferable, adaptable, and protectable.

a) *Memorable*: A mandatory condition for building brand is to create awareness of the brand. Brand elements that garner attention for the service facilitate in brand recall. For example, when thinking of online travel booking "Make My Trip" comes to mind.
b) *Meaningful*: The brand should be able to convey the meaning with the help of information regarding function of the service. It should also highlight the specific attributes of the brand.
c) *Likable*: Brand elements should be able to enrich the image of the brand by making it interesting and fun so that the users may like it.
d) *Transferable*: It means that how much a brand element can add to the equity of a brand. The extent to which it can be useful in line extensions, moving across geographic boundaries and different market segments.
e) *Adaptable*: The fifth factor is the flexibility or adaptability of the brand elements over time. It is easier to update a brand which is flexible.
f) *Protectable*: It means how much the brand element is protectable in the competitive and legal sense. Marketers should choose such

brand elements which are legally protected, formally registered and forcefully defended from unauthorized competition (Keller et al., 2015).

7.4 DMOs STRATEGIES

The DMOs should also *focus on local tourism*, as sometimes these generate 40% more bookings compared with destinations when provided the option of online bookings (Fuggle, 2016).

Destination marketing organizations can use *data analytics*, for example, with the help of Google Analytics, they can gather information about the users like where they are from, which profession do they belong and what their motivations are for visiting a travel webpage (Revfine.com, n.d.).

As majority of the organizations are still following the traditional forms to manage their organizations. There is a need to change this strategy and become *updated*, otherwise, the organization would lose online customers.

7.5 EXPERIENCES STRATEGIES

It is imperative for destination marketers to align customer experiences with customer expectations. This is true for both *online* as well as *offline experiences*. For this, the service providers need to deliver the service with speed and convenience. For example, if a tourist purchases the services online, but requires offline pick-up, then there needs to be a balance between customer expectations (online) and experiences (offline).

In an experience economy, there is a need to *create engagement with experiential marketing*. Several countries have been using such tactics in the form of campaigns to boost tourism, for example, "Amazing Thailand- Always Amazes You" or "Canada- Keep Exploring.. Here, it can be seen that the first part of the slogan focuses on the destination whereas the second part deals with the emotional part in the form of emotions and feelings. In this way, when a tourist visits a destination, he/she remembers the experiences being of the *emotional connection created* by engaging the senses of the travelers.

7.6 DIGITAL AND SOCIAL MEDIA STRATEGIES

Instagram has become a place for opportunities in case of tourism companies. In comparison to Facebook, the number of Instagram users is on the rise at a higher rate. Therefore, marketers need to use this fact to the best of their advantage. The travelers use the *hashtags* set by tourism companies to guide them through the destinations.

Facebook is a platform which inspires travelers to visit locations posted by other travelers. Destination marketers can use this as an opportunity to influence the decision-making of travelers by asking current customers to *post their experiences* so that it inspires other travelers to choose the same destination. The companies can plan *loyalty programs* by rewarding their customers who post-messages in exchange for discounts and perks (Nyorani, 2020).

Twitter is another important social media platform is Twitter. Travel companies need to create a Twitter handle which provides *engaging content* that is *updated regularly*. The travel tips provided here should relate to *personas of the target population*. The marketers should also take care of updating to the latest functionalities provided by the social media platforms.

Websites: The DMOs need to take care that their websites are *not outdated* as they would lose their customer to the competitors. The visual *aesthetics* of the website should be compelling to the customers as they form an impression within the first few seconds of opening the website.

The websites should also be *mobile friendly* as a vast majority of bookings nowadays are made through mobiles. These should *interactively display* text, audio, photos, video, booking systems, and real-time comments from customers.

7.7 SUMMING UP

There is requirement of proper segmentation, targeting, and positioning strategies to market the destinations in such a way that all customers (tourists) are happy. Further, each of seven Ps of service marketing mix plays an important role in providing a memorable experience to the destination visitors. There is also requirement of appropriate criteria for choosing brand elements so that the destination branding leaves a lasting impression on the travelers' minds. Some other kinds of strategies like those related to DMOs, digital, and social media including Facebook, Twitter, Instagram, and official websites, and destination experiences can help destination marketers to market destinations in fruitful ways.

KEYWORDS

- **destination marketing strategies**
- **marketing mix**
- **strategies**
- **branding strategies**
- **experience**
- **strategies**
- **digital and social media strategies**

CHAPTER 8

Sustainability, Challenges, and Future Directions

8.1 SUSTAINABILITY OF DESTINATION MARKETING

There have been environmental changes globally due to human intervention, but not much has been done to move toward a sustainable society. The effects of climate change are visible when there is a consequent rise in the purchase of sustainable goods and services. The term sustainability can be said to be quite old and has been used by most corporates as part of their strategies for long. As such, the younger generation may not relate to the old concept and they may not find it appealing. This may cause information overload for them and may lead to the opposite effect to what was intended.

8.1.1 CHALLENGES IN SUSTAINABILITY

Even though consumers have awareness about the sustainable products, they do not consume sustainable products eventually which lead to wide gap between the two. This is mainly because consumer's awareness and purchase intentions do not convert into actual buying as they are attracted toward other products which are more appealing. Therefore, a major challenge for marketers is to convert sustainable tourism into hedonic tourism by making it more appealing. Care should be taken in such cases that sustainability should not be overemphasized as too much stress on it may lead to "traveler's guilt" which may in turn cause discomfort among travelers (Font and Hindley, 2017).

Destination Marketing: Creating Memorable Tourism Experiences. Rupa Rathee & Pallavi Rajain (Authors)

8.1.2 OPPORTUNITIES OF SUSTAINABILITY

In order to gain commercial advantage in tourism, it is needed to be integrated with marketing as tourism industry is increasing in terms of economic value in a competitive world. But marketing selective destinations which are friendly to the environment in order to attract sustainable tourists is not economical as it ignores tourists who do not have such an inclination. Therefore, the aim of marketers should be to maximize the number of customers by persuading them to buy more sustainable products.

8.1.3 SUSTAINABILITY IN EXPERIENCE ECONOMY

With the emergence of experience economy there has been a drastic change in the way in which products are marketed. This poses a challenge for destination marketing organizations, which need to adapt by providing distinct and memorable experiences to customers. These emotional experiences can be created by appealing to the senses of the customers. For example, allowing the tourist to taste the local cuisine of the place and the smell of fresh air can be source of sustainable yet memorable experiences. The travel to mountainous, hilly or beach areas is another source of pleasure for the tourists along with being sustainable. Such hilly areas also pose ban on products that can be harmful to the environment.

For example, in India, Uttarakhand government (major state with hill stations) has imposed a strict ban since August 2018, on the usage of polythene bags in its bid to become a plastic-free state. The Sikkim government has also posed ban on mineral water bottles as tourists generally leave them lying around. Instead, they provide other safe options.

Another example is of Victoria Memorial which has banned plastic bags (Figure 8.1).

Care should be taken when projecting sustainable goods and services through advertisements so that there may be no negative impact of these communication strategies which may lead to bad experiences. It has been found in various studies that advertisements which focus more on the experiences provided by the destination rather than only sustainability factors are preferred more as people do not prefer to travel just for the sake of sustainability. For example, when advertising a destination in the countryside, it would be better to focus on the attractions and highlights of the place rather than simply a ride on the bicycle.

FIGURE 8.1 Sikkim tourism environmental concerns.
Source: © WWF-India

8.1.4 SHIFT TOWARD SUSTAINABILITY

The shift towards sustainability should be gradual as marketers are constantly on the lookout as how to communicate and share knowledge of the sustainability of tourism. This can be done effectively by using visuals overtly and covertly for projecting the sustainability of the tourist destination. Studies have suggested limited persuasiveness of overt messages of sustainability. In a study by Hanna et al. (2018), one of the interviewees said in the focus group interview that -

"I think they need to sell the place before talking about sustainability because sustainability is an extra incentive, thinking "wow it looks great AND it's this (sustainable)' rather than 'it's sustainable!'"

Studies have shown that organizations that market sustainability by showing themselves as green may at times appear to be too nerdy. Therefore, the shortcomings of sustainability as an old concept may be improved by turning them into something new.

For example, electric vehicles, green labeling, energy star ratings and organic foods are some of the sustainable items which appear new and appealing to customers.

Besides these, certain events are also labeled as green events. Such events reduce the use of paper by using digital material, thus reducing the carbon footprint.

In case of services, an example of sustainable destination in India, The Ramoji film city, Hyderabad has the largest complex in the world but does not allow personal conveyance to enter instead they have provided their own buses for travel in the premises. The tourists can use these services to enjoy their time along with protecting the environment as a part of the sustainability initiative by the destination (Figure 8.2).

Another example is of Taj Mahal where Supreme Court order restricts any vehicle run by fuel to enter within 500 m of the monument. Even US President Trump during his visit to the Taj in February 2020 had to replace his cavalcade with electric buses and battery run golf carts (Figure 8.3).

The strategy which embeds the concepts of sustainability within the destination marketing theme would be best in today's time when there is a need to conserve and protect the natural environments. In order to increase tourist's engagement toward sustainable destinations, there is a need for dissemination information through indirect methods like videos which hold great persuasion power. Increasing the engagement of tourists with sustainable tourism products while traveling can be possible when sustainable tourism marketing is done with full force.

FIGURE 8.2 Ramoji Film City.

Source: Photo by Vinayaraj. https://creativecommons.org/licenses/by-sa/3.0/deed.en

FIGURE 8.3 Battery powered vehicle in Taj Mahal, Agra.
Source: Photo by Isriya Paireepairit. https://creativecommons.org/licenses/by-nc/2.0/

8.2 CHALLENGES IN DESTINATION MARKETING

There are mainly two forms of challenges in the field of destination marketing: marketing challenges and research challenges.

Marketing challenges arise due to certain societies being against tourism for various reasons like:

a) *Disliking strangers*: Certain communities although small in number do not welcome strangers for example certain parts of Hawaiian Islands.
b) *Changing destination characteristics*: The characteristics and behavior of visitors may adversely affect the destination characteristics.
c) *Negative impact on society*: The rise in tourism may be accompanied by rise in negative activities like thefts and other crimes.
d) *Damage to economy*: Some local residents provide an argument that the jobs related to tourist activities are low-paying and the high-paying jobs only result in out flow of profits outside the destination.
e) *Hazard to environment*: Tourist activities may also lead to detrimental effects on the environment.
f) *Outside control*: Local residents believe that tourism gives control to outsiders in the form of colonialism.

8.2.1 CHALLENGES FOR DESTINATION MARKETERS

Catering to niche groups: Destination marketers face the challenge of providing facilities to a specific group. For example, providing accommodation in Hawaii to an adult-only population. Hawaii is best known for its beaches has a crowded cluster of hotels for families. In order to cater to an adult-only population, to provide a kid's free environment became a challenge for hotels and resorts. This problem was overcome by providing personalized and unique experiences.

Customer experience enhancement: Providing comfortable means of transportation has also been a challenge. In order to provide comfort to passengers, transportation services need to prevent overcrowding but this leads to lower profits. Thus, providing transportation either through railways, airways or roadways which is both profitable to service provider and comfortable to the passengers has been a challenge. In order to overcome this challenge, airlines nowadays provide extra facilities like more leg space and variety in food served at additional costs.

Choosing from alternatives: Nowadays, travelers have a lot of options to gather travel information either through online booking websites or travel blogs. There is an endless list of options for the people to choose from. This makes it difficult for travel agents to grab the attention of travelers so that they have an experiential travel. This challenge can be overcome by producing vlogs or video logs which can showcase travel experience to customers in a crowded market.

Shifting markets: There has been a huge shift in the travel market. Around a decade ago, very few reservations were made online. Compared with that, 60% of the bookings now made are online. Those tour operators who could not change along with this shift did not survive the market.

Slump in off-seasons: Every tourist destination has a peak season and a slump season. The retail and shopping centers located within the tourist destinations face the challenge during these slumps with dipping sales. They need to have a strategy in order to prepare for the slump in off-seasons. For example, the souvenir shops can sell their products online during off -seasons. Another option is to connect with the audience using contextual marketing campaigns.

Staying top of the game: When tourists are flooded with options then staying on top of the game is not easy. In case of various restaurants and eateries, this has been a major challenge. They use various promotion techniques and provide discounts in order to attract customers. The marketers can also widen their digital footprint in order to gain more visibility. This can be done by asking happy customers to leave reviews online which can help to build trust.

Attracting different people: Attracting tourists of different demographics has been a challenge for certain tourist spots. For example, Museums generally attract history enthusiasts whereas gardens and parks are for lovers of nature. With the lack of focus on attracting different demographics, these sites face loss in sales. Disney world initially catered to only small children and families. But later, it evolved into multitheme park with after hour dining, shopping, and entertainment (Finch, 2019).

Pandemic: Due to several travel restrictions imposed on tourists to keep them safe, a pandemic situation poses huge challenge for destination marketers. Most sites like amusement parks, museums, sports venues remain closed. For example, during the spread of SARS-Cov2 (COVID-19) in 2020, governments around the world, imposed travel restrictions and lockdowns. This deterred the tourism industry to perform its functions as this service industry mainly works on traveling activities of customers. The restrictions and lockdowns resulted in job losses due to loss in revenue for airlines, travel agents, hotel, and restaurant employees.

8.2.2 RESEARCH CHALLENGES IN DESTINATION MARKETING

i) *Scale*: The first and foremost issue for researchers is the development of scale. This further can be divided into two tasks, that is, development of scale based on experiences of tourists and scale based on how researchers conceptualize various phenomena related to destination marketing. Even though some researchers have focused on the need for a combined system for scale development, but still there is need to understand the different scales being used at present having focus on different types of destination environments.

ii) *Approach*: Various approaches are considered when understanding destination marketing experiences. It may be seen in some studies that authors have used quantitative techniques to determine destination marketing activities through non-personalized methods which may further be detailed through phenomenological studies. Besides these epistemological studies is another form of studying the exploration of destination experiences.

iii) *Interactions*: Several new technologies have come up in recent times which make interaction between consumers and producers more viable. These technologies help to keep track of tourists' interests and movements through "big data" and other data generating technologies. However, there are some ethical concerns regarding use of such technologies and they lead to infringement of privacy of the customers (Lugosi and Walls, 2013).

8.2.3 GAP IN THEORY AND PRACTICE

In an ideal society, the theories developed by researchers would provide insights to industry and the best practices developed by industry would add to the literature, thus benefitting both. However, there exist a gap between theory and practice. There have been many works related to destination marketing under various terms like hospitality, leisure, travel, and tourism. Even though all these works are inter-related, one should not assume that researchers work in an organized fashion to achieve a common target. Academicians are threatened by the phrase "publish or perish," so they may work under this pressure without actually understanding the need for the research. On the other hand, in the tourism industry most of the businesses are small and thus focused only on their day-to-day operations, without giving a thought to what knowledge is being shared by the academicians (MacCannell, 1976).

In the future, however, the scenario may be different a lot of degree courses related to hospitality have come up since the 1990s. The future tourism managers and destination marketers would have had ample exposure to tourism theories, thus reducing the gap that exists.

8.3 FUTURE OF DESTINATION MARKETING

Advancements in technology have been transforming every industry. This is no different for tourism industry as well. Digital marketing tools have transformed marketing in the tourism industry too. Digital technologies, such as artificial intelligence, robots, virtual reality have opened huge number of possibilities for innovations in marketing.

Some of these options to be used in the future include:

Tracking technology: Tracking movements of tourists using GPS helps destination marketers to understand their behaviors and motives (Mckercher et al., 2019) (Figure 8.4).

Integrating VR devices: Using VR head mounted devices have been found to provide more engaging and immersive experiences in comparison to mobile phones and laptops (Flavián et al., 2019) (Figure 8.5).

Aerial drone videos: Visually appealing videos of destinations generated through the use of drones seem to have been attracting much attention. The YouTube videos captured through drones provide some useful content regarding the destination being captured (Stankov et al., 2019) (Figure 8.6).

Google Earth: Google provides a bundle of services which will be tremendously useful for tourists in planning their travel to various destinations. These include:

a) *Google Earth* which is an application related to Google Maps that uses satellite photography to display multiple maps. It was initially developed under the name of Earthviewer 3D by the company Keyhole Inc and funded by Central Intelligence Agency. It was later acquired by Google. This application makes use of aerial photographs, satellite images, and geographic information which can be accessed by laptops, mobiles, and tablets. Such images are useful for travelers who want to explore a destination before actually visiting it (Rosselló, 2017) (Figure 8.7).

FIGURE 8.4 Tracking technology.
Source: Illustration by Wiki0717. https://creativecommons.org/licenses/by-sa/4.0/deed.en

FIGURE 8.5 VR devices.

FIGURE 8.6 Aerial drone videos.

Source: Photo by miguelb. https://creativecommons.org/licenses/by/2.0/

FIGURE 8.7 Google Earth.

Source: Google Earth.

b) *Google Street View* is a feature which is also related to Google Earth providing a panoramic view of the 360 degree. It allows users to look at various cities around the world and their surroundings.
c) *Google Assistant* provides information using predictions based on user's search history. It helps user to locate nearby attractions, events and can have two-way conversations with the users in over 30 languages.

d) *Google flights* help users to book flights through different suppliers. It has several features like suggesting various destination options according to the tastes of the users depending on the budget and time suggested by the traveler.

Marketing robot services: Humanoid robot or human-like robots have started to become part of the tourism and hospitality workforce. They can in future play roles as receptionist, waiter, and tour guides. This is mainly because of better memory recall and proficiency in many languages (Murphy et al., 2019) (Figures 8.8 and 8.9).

FIGURE 8.8 Humanoid service.

Source: Photo by Jasonprost. https://creativecommons.org/licenses/by-sa/4.0/deed.en

Space tourism: It is "human space travel for recreational purposes" (Dunk, 2011). It may involve travel around the earth's orbit or to the moon. Till date, only the Russian Space Agency has been able to perform the orbital space tourism. Companies, such as SpaceX, Virgin Galactic, and Blue Origin are involved in the process of developing space tourism vehicles. NASA has been planning moon tourism for quite some time. It had announced in 2019 that with the help of SpaceX and Boeing it would allow private astronauts to the International Space Station. Space Adventures, a space tourism company has been working on taking travelers to the moon with price per passenger being $100,000,000.

FIGURE 8.9 Humanoid receptionist.

Health and safety: This is another important aspect that needs to be looked into during future travel activities. The health and safety measures came into highlight as a result of the spread of COVID-19 in 2020. The travel industry needs to be prepared to deal with such situations to maintain their customer base. For example, a restaurant in Amsterdam built small glass cubicles so that customers could enjoy their meals while social distancing (Humphries, 2020) (Figure 8.11).

Another example is of Disneyland where enhanced safety measures include temperature checks at entry points, addition of hand-washing stations, physical distancing by reduction in park capacity and cashless transactions (Figure 8.12).

FIGURE 8.10 Dining while social distancing.

FIGURE 8.11 Disneyland before and after Covid-19.

A food outlet in Bengaluru, India provides contactless dining in their premises (Figure 8.13).

FIGURE 8.12 Contactless dining at a food outlet in Bengaluru.

After the vaccination process started in 2021, stringent measures were taken by government of countries all over the world to restrict travel on the basis of production of vaccination certificates. This helped in the revival of tourism which had dropped by 80% in 2020 after Covid-19. Travelers and marketers felt safe to reopen the various tourist spots which were under lockdown for many months (ADB Briefs, August 2021).

Various countries made different efforts to recapture international travelers. One example is of Thailand, they used the "Sandbox" approach to attract travelers. Through this initiative fully vaccinated travelers were offered exemption from quarantine to travel in a particular city for example Phuket. Travelers were provided accommodation in establishments certified by Safety & Health Administration of the Thai government. Such models helped in the reopening of tourist destinations throughout Thailand (Saxon et al., 2021).

The hospitality industry especially in India looks forward to the government that they may incentivize domestic travel through tax cuts so that local tourism may survive until international travel resumes. People are looking forward to travel to beat the fatigue of restrictions due to the pandemic and it is a good time for the government to "hard sell India" as a destination with facilitative measures and incentives to encourage the million plus people who travel out to stay back and holiday within the country (PTI, 2021).

8.4 SUMMING UP

Even though consumers have awareness about the sustainable products, they do not consume sustainable products eventually which lead to wide gap between the two. Therefore, a major challenge for marketers is to convert sustainable tourism into hedonic tourism by making it more appealing. Increasing the engagement of tourists with sustainable tourism products while traveling can be possible when sustainable tourism marketing is done with full force. Further, there are two forms of challenges in the field of destination marketing: marketing challenges and research challenges. However, there exist a gap between theory and practice. The future tourism managers and destination marketers would have had ample exposure to tourism theories, thus reducing the gap that exists. Digital technologies like artificial intelligence, robots, virtual reality have opened huge number of possibilities for innovations in destination marketing.

KEYWORDS

- **sustainability**
- **challenges**
- **future**
- **VR devices**
- **space tourism**

References

Aaker, D. A. *Capitalizing on the Value of a Brand Name*; Free Press: New York, 1991a; pp 28, 35–37.

Aaker, D. A. *Managing Brand Equity*; Free Press: New York, 1991b.

Ackers, M. Viral Marketing for Tourism, 2010. https://www.tourismindustryblog.co.nz/2010/03/viral-marketing-for-tourism/#:~:text=It%20will%20be%20targeted%20at,of%20the%20brand%20or%20message).&text=A%20couple%20of%20current%20examples,Zealand%20%E2%80%93%20Who%20Would%20You%20Spoon%3F

ADB Briefs. Can Vaccination Help Restart Tourism? 2021

Akehurst, G.; Bland, N.; Nevin, M. Tourism Policies in the European Community Member States. *Int. J. Hosp. Manage.* **1993,** *12* (1), 33–66.

Alhroot, A. H. J. *Marketing of a Destination: Jordan as a Case Study*; Doctoral thesis, University of Huddersfield, 2007.

Alizadeh, A.; Isa, R. M. The Use of Social Media in Destination Marketing: An Exploratory Study. *Turizam: međunarodniznanstveno-stručničasopis* **2015,** *63* (2), 175–192.

Antoniadis, K.; Zafiropoulos, K.; Vrana, V. Communities of Followers in Tourism Twitter Accounts of European Countries. *Eur J. Tour. Hosp. Recreat.* **2015,** *6* (1), 11–26.

Atorough, P.; Martin, A. The Politics of Destination Marketing: Assessing Stakeholder\ Interaction Choice Orientations toward a DMO Formation, Using the Thomas-Kilmann Conflict\Mode Instrument. *J. Place Manage. Dev.* **2012,** *5* (1), 35–55.

Barnes, S. J.; Mattsson, J.; Sørensen, F. Destination Brand Experience and Visitor Behavior: Testing a Scale in the Tourism Context. *Ann. Tour. Res.* **2014,** *48*, 121–139.

Belch, G.; Belch, M. A. *Advertising and Promotion–an Integrated Marketing Communications Perspective*, Vol. 5; Auflage: New York, 2001.

Bladen, C.; Kennell, J.; Abson, E.; Wilde, N. *Events Management-An Introduction*; Routledge: London, 2012.

Bloggers Are in All Praise of Kerala. Retrieved September 25, **2018,** from DECCAN CHRONICLE. https://www.deccanchronicle.com/nation/current-affairs250318bloggers-are-in-all-praise-of-kerala.html

Bokunewicz, J. F.; Shulman, J. Influencer Identification in Twitter Networks of Destination Marketing Organizations. *J. Hosp. Tour. Technol.* **2017,** *8* (2), 205–219.

Bratman, M. E.; Israel, D. J.; Pollack, M. E. Plans and Resource–Bounded Practical Reasoning. *Comput. Intell.* **1988,** *4* (3), 349–355.

Buhalis, D. Marketing the Competitive Destination of the Future. *Tour. Manage.* **2000,** *21* (1), 97–116.

Carter, L. Destination Branding: The Key Essentials for Success, 2014. https://www.personadesign.ie/destination-branding-the-key-essentials-for-success/

Chandra, S.; Menezes, D. Applications of Multivariate Analysis in International Tourism Research: The Marketing Strategy Perspective of NTOs. *J. Econ. Soc. Res.* **2001,** *3* (1), 77–98.

Che, D.; Veeck, A.; Veeck, G. Sustaining Production and Strengthening the Agritourism Product: Linkages among Michigan Agritourism Destinations. *Agric. Hum. Values* **2005,** *22* (2), 225–234.
Chen, H. T.; Lin, T. W. How a 3D Tour Itinerary Promotion Affect Consumers' Intention to Purchase a Tour Product? *Info. Technol. J.* **2012,** *11* (10), 1357.
Chohan, S. Download Rajasthan Tourism Ad Songs & Lyrics, 2016. https://www.udaipurblog.com/download-rajasthan-tourism-ad-songs-lyrics.html
Choudhary, R. R.; Dixit, S. K. Rural Tourism in India: Exploring the Travel Blogger's Perspective, Rurality. *Rural. Rural Tour.* **2019,** 51–64.
Coban, O. Event Tourism. In *Global Issues and Trends in Tourism*; St. KlimentOhridski University Press, 2016.
Commonwealth of Australia. Towards a National Sports Tourism Strategy (PDF), 2000. Retrieved August 11, 2020.
Crouch, G. I.; Ritchie, J. R. B. Tourism, Competitiveness, and Social Prosperity. *J. Busi. Res.* **1999,** *44*, 137–152.
Di Pietro, L.; Di Virgilio, F.; Pantano, E. Social Network for the Choice of Tourist Destination: Attitude and Behavioural Intention. *J. Hosp. Tour. Technol.* **2012,** *3* (1), 60–76.
Engel, J. F.; Kollat, D. T.; Blackwell, R. D. *Consumer Behavior*, 1st ed.; Holt, Rinehart and Winston: New York, 1968.
Enright, M. J.; Newton, J. Tourism Destination Competitiveness: A Quantitative Approach. *Tourism Management* **2004,** *25* (6), 777–788.
Fabricius, M. P. Competitive Strategy for Tourism Destination; Unpublished MBL Research Report, University of South Africa, Pretoria, 2001.
Finch, S. 7 Tourism Marketing Challenges and How to Overcome Them, 2019. https://marketing.sfgate.com/blog/tourism-marketing-challenges
Flavián, C.; Ibáñez-Sánchez, S.; Orús, C. Integrating Virtual Reality Devices into the Body: Effects of Technological Embodiment on Customer Engagement and Behavioral Intentions toward the Destination. *J. Travel Tour. Market.* **2019,** *36* (7), 847–863.
Font, X.; Hindley, A. Understanding Tourists' Reactance to the Threat of a Loss of Freedom to Travel Due to Climate Change: A New Alternative Approach to Encouraging Nuanced Behavioural Change. *J. Sustain. Tour.* **2017,** *25* (1), 26–42.
Frian, K. Embracing Agritourism. *J. Busi.* **2010,** *25* (16), B11.
Fuggle, L. The 8 Destination Marketing Strategies with the Highest ROI, 2016. https://www.trekksoft.com/en/blog/best-destination-marketing-strategies
Gammon, S.; Robinson, T. Sport and Tourism: A Conceptual Framework. *J. Sport Tour.* **2003,** *8* (1), 21–26. doi:10.1080/14775080306236.
Garay, L.; Morales Pérez, S. Understanding the Creation of Destination Images through a Festival's Twitter Conversation. *Int. J. Event Fest. Manage.* **2017,** *8* (1), 39–54.
Gartner, W. C. Image Formation Process. *J. Travel Tour. Market.* **1993,** *2* (2/3), 191, 215.
Getz, D. *Event Management and Event Tourism*, 2nd ed.; Cognizant: New York, 2005.
Getz, D. *Event Studies. Theory, Research and Policy for Planned Events*; Butterworth-Heinemann: Oxford, 2007.
Getz, D. Event Tourism: Definition, Evolution and Research. *Tour. Manage.* **2008,** *29* (3), 403–428.

Ghazali, R. M.; Cai, L. Social Media Sites in Destination Image Formation. In *Tourism Social Media: Transformations in Identity, Community and Culture*; Emerald Group Publishing Limited, 2014; pp 73–86.

Gretzel, U.; Yoo, K. H. Premises and Promises of Social Media Marketing in Tourism. *Routledge Handbook Tour. Market.* **2013,** 491–504.

Grewal, D.; Levy, M. *Marketing Management*; Pearson, 2018.

Gtitravel. How Is Business Travel Different from Personal Travel, 2017. https://www.gtitravel.com/how-is-business-travel-different-from-personal-travel/

Guitart, J.; Torres, J.; Ayguadé, E. A Survey on Performance Management for Internet Applications. *Concurren. Comput. Pract. Exp.* **2010,** *22* (1), 68–106.

Hamill, J.; Stevenson, A.; Attard, D. National DMOs and Web 2.0. *Soc. Media Travel Tour. Hosp. Theory Pract. Cases* **2012,** 97–120.

Hammitt, W. E.; Bixler, R. D.; Noe, F. P. Going beyond Importance-Performance Analysis to Analyze the Observance-Influence of Park Impacts. *J. Park Recreat. Admin.* **1996,** *14* (1), 45–62.

Hanna, P.; Font, X.; Scarles, C.; Weeden, C.; Harrison, C. Tourist Destination Marketing: From Sustainability Myopia to Memorable Experiences. *J. Destin. Market. Manage.* **2018,** *9*, 36–43.

Hartl, A. Developing Marketing Strategies for Tourism Destinations in Peripheral Areas of Europe; Doctoral dissertation, Bournemouth University, 2002.

Hasan, M.; Jobaid, M. I. Heritage Tourism Marketing: Status, Prospects and Barriers. *IOSR J. Busi. Manage.IOSR-JBM)* **2014,** *16* (5), 40–48.

Heath, E.; Wall, G. *Marketing Tourism Destinations: A Strategic Planning Approach*; Wiley: New York, 1992.

Holbrook, M. B. Consumption Experience, Customer Value, and Subjective Personal Introspection: An Illustrative Photographic Essay. *J. Busi. Res.* **2006,** *59* (6), 714–725.

Holcolm, B. Marketing Cities for Tourism. In *The Tourist City*; Judd, D. R., Fainstein, S. S., Eds.; Yale University Press: Newhaven, 1999; pp 54–70.

Horowitz, M. D.; Rosensweig, J. A.; Jones, C. A. Medical Tourism: Globalization of the Healthcare Marketplace. *MedGenMed.* **2007,** *9* (4), 33. PMC 2234298. PMID 18311383.

Howard, J. A. *Marketing Management: Analysis and Planning*; Irwin: Homewood, ILL, 1963.

Howard, J. A.; Sheth, J. N. *The Theory of Buyer Behavior*; New York, 1969; p 63.

Hu, Y.; Ritchie, J. R. B. Measuring Destination Attractiveness: A Contextual Approach. *J. Travel Res.* **1993,** *32* (2), 25–34.

Hudson, S.; Ritchie, J. B. Branding a Memorable Destination Experience. The Case of 'Brand Canada'. *Int. J. Tour. Res.* **2009,** *11* (2), 217–228.

Humphries, M. A Restaurant in Amsterdam Built Small Greenhouses for Diners to Enjoy a Meal in While Social Distancing, 2020. https://www.insider.com/amsterdam-restaurant-mini-greenhouses-cubicles-social-distance-2020-5

Inkson, K.; Kolb, D. *Management*; Addison Wesley Longman: Auckland, 1998.

Johnson, G. P. *Conference Marketing*, n.d. https://www.gpj.com/worldwide-experience-marketing/conference-marketing/#:~:text=meet%20your%20challenge.-, Conference%20 Management%20Services,others%20together%20under%20one%20umbrella.

Kavoura, A.; Stavrianeas, A. The Importance of Social Media on Holiday Visitors' Choices-the Case of Athens, Greece. *EuroMed J. Busi.* **2015,** *10* (3), 360.

Keller, K. L. *Strategic Brand Management-Building, Managing, and Measuring Brand Equity*, 4th ed.; Pearson, 2013.

Keller, K. L.; Parameswaran, A. M. G.; Jacob, I. *Strategic Brand Management-Building, Measuring, and Managing, Brand Equity*, 4th ed.; Pearson, 2015.

Kim, J. H.; Ritchie, J. B. Cross-Cultural Validation of a Memorable Tourism Experience Scale (MTES). *J. Travel Res.* **2014,** *53* (3), 323–335.

Koo, C.; Shin, S.; Gretzel, U.; Hunter, W. C.; Chung, N. Conceptualization of Smart Tourism Destination Competitiveness. *Asia Pacific J. Info. Syst.* **2016,** *26* (4), 561–576.

Kotler, P.; Bowen, J. T.; Makens, J.; Baloglu, S. *Marketing for Hospitality and Tourism*; Pearson, 2016.

Kotler, P.; Keller, K. L.; Ang, S. H.; Tan, C. T.; Leong, S. M. *Marketing Management: An Asian Perspective*; Pearson, 2018.

Leung, D.; Law, R.; Van Hoof, H.; Buhalis, D. Social Media in Tourism and Hospitality: A Literature Review. *J. Travel Tour. Market* **2013,** *30* (1–2), 3–22.

Leung, X. Y.; Jiang, L. How Do Destination Facebook Pages Work? An Extended TPB Model of Fans' Visit Intention. *J. Hosp. Tour. Technol.* **2018,** *9* (3), 397–416.

Line, N. D.; Wang, Y. A Multi-Stakeholder Market-Oriented Approach to Destination Marketing. *J. Destin. Market. Manage.* **2017,** *6* (1), 84–93.

Lugosi, P.; Walls, A. R. Researching Destination Experiences: Themes, Perspectives and Challenges. *J. Destin. Market. Manage.* **2013,** *2* (2), 51–58.

Lugosi, P.; Walls, A. R. Researching Destination Experiences: Themes, Perspectives and Challenges. *J. Destin. Market. Manage.* **2013,** *2* (2), 51–58.

MacCannell, D. *The Tourist*; Shocken Books: New York, 1976.

Marasco, A.; Buonincontri, P.; van Niekerk, M.; Orlowski, M.; Okumus, F. Exploring the Role of Next-Generation Virtual Technologies in Destination Marketing. *J. Destin. Market. Manage.* **2018,** *9*, 138–148.

Marcussen, C. H. The Effects of Internet Distribution of Travel and Tourism Services on the Marketing Mix: No-Frills, Fair Fares, and Fare Wars in the Air. *Info. Technol. Tour.* **1999,** *2* (3), 197–212.

Martilla, J. A.; James, J. C. Importance-Performance Analysis. *J. Market.* **1977,** *41* (1), 77–79.

Mattsson, J.; Sundbo, J.; Fussing-Jensen, C. Innovation Systems in Tourism: The Roles of Attractors and Scene-Takers. *Indust. Innov.* **2005,** *12* (3), 357–381.

Mayo, E. J.; Jarvis, L. P. *The Psychology of Leisure Travel. Effective Marketing and Selling of Travel Services*; CBI Publishing Company, Inc, 1981.

Mckercher, B.; Hardy, A.; Aryal, J. Using Tracking Technology to Improve Marketing: Insights from a Historic Town in Tasmania, Australia. *J. Travel Tour. Market.* **2019,** *36* (7), 823–834.

Mei, T.; Hua, X. S.; Yang, L.; Li, S. VideoSense: Towards Effective Online Video Advertising. In *Proceedings of the 15th ACM International Conference on Multimedia*; ACM, Sept 2007; pp 1075–1084.

Mendiratta, A. Evaluating and Evolving the Role of the Destination Marketing Organization, 2014. https://destinationsinternational.org/evaluating-and-evolving-role-destination-marketing-organization

Morrison, A.; Anderson, D. Destination Branding. In *Missouri Association of Convention & Visitor Bureaus Annual Meeting*, June 2002.

Murphy, J.; Gretzel, U.; Pesonen, J. Marketing Robot Services in Hospitality and Tourism: The Role of Anthropomorphism. *J. Travel Tour. Market.* **2019,** *36* (7), 784–795.

Myers, J. H. Positioning Products/Services in Attitude Space. *Market. Res.* Mar **1992,** 46–51.

National Trust for Historic Preservation. Heritage Tourism, 2008. www.nationaltrust.org/heritage_tourism

Neuhofer, B.; Buhalis, D.; Ladkin, A. Conceptualising Technology Enhanced Destination Experiences. *J. Destin. Market. Manag.* **2012,** *1* (1–2), 36–46.

Ng, B. The 4 + 3 P's in the Marketing Mix for Tour Operators, 2018. https://www.rezdy.com/blog/the-4-3-ps-in-the-marketing-mix-for-tour-operator/

Nyorani, C. How to Use Social Media for Destination Marketing, 2020. https://pro.regiondo.com/destination-marketing-social-media/

NZTPD. 75 Years of Tourism; New Zealand Tourist & Publicity Department: Wellington, 1976.

Oh, H. Revisiting Importance–Performance Analysis. *Tour. Manage.* **2001,** *22* (6), 617–627.

Ontiveros, J. The Impact of Travel Blogs in Destination Marketing, 2016. http://www.brandba.se/blog/2016/11/10/the-impact-of-travel-blogs-in-destination-marketing#:~:text=People%20can%20share%20blogs%20posts,new%20visitors%20to%20the%20destination.&text=Travel%20Blogs%20have%20shown%20to,market%20the%20destinations%20among%20people

Oppermann, M. Destination Threshold Potential and the Law of Repeat Visitation. *J. Travel Res.* **1998,** *37* (2), 131–137.

Packer, J.; Ballantyne, R. Conceptualizing the Visitor Experience: A Review of Literature and Development of a Multifaceted Model. *Visitor Stud* **2016,** *19* (2), 128–143.

PATA. The Rise of Stopover Tourism, 2019. http://patachina.cn/ziyuan/?filename=/upload/2019/05/PATA-VE-Bulletin-January-2019-n.pdf

Pike, S. *Destination Marketing Organizations*; Routledge: London, 2004.

Pike, S. Spoilt for Choice—Short Break Holiday Preferences in the Brisbane Market. The Council for Australian University Tourism and Hospitality Education (CAUTHE) Annual Conference; University of Queensland: Brisbane, Feb **2004**; pp 577–586.

Pike, S. *Destination Marketing-An Integrated Marketing Communication Approach*; Routledge: London, 2007.

Pine, B. J.; Gilmore, J. H. *The Experience Economy: Work Is Theatre & Every Business a Stage*; Harvard Business Press: Cambridge, 1999.

Pouder, R. W.; Clark, J. D.; Fenich, G. G. An Exploratory Study of How Destination Marketing Organizations Pursue the Sports Tourism Market. *J. Destin. Market. Manage.* **2018,** *9*, 184–193.

Prahalad, C. K.; Ramaswamy, V. Co-Creation Experiences: The Next Practice in Value Creation. *J. Interact. Market.* **2004,** *18* (3), 5–14.

PTI. Survival first, revival later: Hospitality industry's mantra for 2022. 2021. Retrieved from https://www.business-standard.com/article/companies/survival-first-revival-later-hospitality-industry-s-mantra-for-2022-121122000291_1.html

Rathee, R.; Rajain, P. Experiential Marketing: Strategies for the Leisure Industry. In *Breaking Down Language and Cultural Barriers through Contemporary Global Marketing Strategies*. IGI Global, 2019; pp 113–127.

Revfine.com. 14 Destination Marketing Strategies to Attract More Visitors, n.d. https://www.revfine.com/destination-marketing/

Rezzcomm.com. 5 Creative Airport Destination Marketing Campaigns, 2018. https://www.rezzcomm.com/blog/2018/10/25/creative-destination-marketing-campaigns/

Rilla, E.; Hardesty, S.; Getz, C.; George, H. California Agritourism Operations and Their Economic Potential Are Growing. *Calif. Agric.* **2011,** *65* (2), 57–65.

Rosselló, L. G. Tourism Intermediation and the Role of Google, 2017.

Ruddy, J.; Flanagan, S. *Tourism Destination Marketing: Gaining the Competitive Edge.* Conference Proceedings. European Conference of the Travel and Tourism Research Association, 1999.

Saqib, N. A Positioning Strategy for a Tourist Destination, Based on Analysis of Customers' Perceptions and Satisfactions. *J. Tour. Analy. Revista de AnálisisTurístico* **2019**.

Saxon, S.; Sodprasert, J.; Sucharitakul, V. Reimagining Travel: Thailand Tourism After the COVID-19 Pandemic. 2021. Retrieved from https://www.mckinsey.com/industries/travel-logistics-and-infrastructure/our-insights/reimagining-travel-thailand-tourism-after-the-covid-19-pandemic

Schroeder, R. Defining Virtual Worlds and Virtual Environments. *J. Virtual Worlds Res.* **2008,** *1* (1).

Sevin, E. Places Going Viral: Twitter Usage Patterns in Destination Marketing and Place Branding. *J. Place Manage. Dev.* **2013,** *6* (3), 227–239.

Sharda, N.; Ponnada, M. Tourism Blog Visualizer for Better Tour Planning. *J. Vacation Market.* **2008,** *14* (2) 157–167. http://jvm.sagepub.com/content/14/2/157

Shen, H.; Song, C.; Li, M.; Jiang, Q. Shaping Destination Images through Social Networking Sites: A Case Study of Singapore. In *Advances in Hospitality and Leisure*; Emerald Group Publishing Limited, 2015; pp 89–107.

Sherwood, P. How Travel Companies Can Nail Market Segmentation, 2018. https://www.gravitatedesign.com/blog/tourism-market-segmentation-101/#:~:text=Marketing%20segmentation%20in%20the%20travel%20and%20tourism%20industry&text=Effective%20segmentation%20is%20based%20on,behavioral%20patterns%2C%20or%20cognition%20ratings

Sigala, M.; Christou, E.; Gretzel, U., Eds. *Social Media in Travel, Tourism and Hospitality: Theory, Practice and Cases*; Ashgate Publishing, Ltd, 2012.

Sood, S. 10 Sites of Religious Pilgrimage, 2012. http://www.bbc.com/travel/story/20120119-travelwise-10-sites-of-religious-pilgrimage

Soteriades, M. Tourism Destination Marketing: Approaches Improving Effectiveness and Efficiency. *J. Hosp. Tour. Technol.* **2012,** *3* (2), 107–120.

Srikatanyoo, N.; Campiranon, K. Agritourist Needs and Motivations: The Chiang Mai Case. *J. Travel Tour. Market.* **2010,** *27* (2), 166–178.

Stankov, U.; Kennell, J.; Morrison, A. M.; Vujičić, M. D. The View from Above: The Relevance of Shared Aerial Drone Videos for Destination Marketing. *J. Travel Tour. Market.* **2019,** *36* (7), 808–822.

Statista.com Leading Countries Based on Number of Twitter Users as of April 2020, 2020. https://www.statista.com/statistics/242606/number-of-active-twitter-users-in-selected-countries/

Statista.com Number of Monthly Active Twitter Users Worldwide from 1st Quarter 2010 to 1st Quarter 2019, 2020. https://www.statista.com/statistics/282087/number-of-monthly-active-twitter-users/

Tasci, A. D.; Gartner, W. C. Destination Image and Its Functional Relationships. *J. Travel Res.* **2007,** *45* (4), 413–425.

Top Culinary Tourism Destinations, 2016. https://coursefinders.com/blog/en/4168/top-culinary-tourism-destinations

Tsiotsou, R. H.; Goldsmith, R. E. Target Marketing and Its Application to Tourism. In *Strategic Marketing in Tourism Services*; Emerald Group Publishing Limited, 2012.

Tuten, T. L.; Solomon, M. R.; Pilotti, L.; Tedeschi-Toschi, A. *Social Media Marketing: Post-consumo, innovazionecollaborativa e valorecondiviso*; Pearson, 2014.

Ukessays. Tourism Marketing Mix Essay, 2016. https://www.ukessays.com/essays/tourism/marketing-mix-of-tourism-industry-essay.php

Vivion, N. Smart Tourism: Should Destinations Take Ownership of Enhancing the in-Destination Experience through Technology? 2018. https://www.phocuswire.com/Smart-tourism-Should-destinations-take-ownership-of-enhancing-the-in-destination-experience-through-technology

Volo, S. Bloggers' Reported Tourist Experiences: Their Utility as a Tourism Data Source and Their Effect on Prospective Tourists. *J. Vacation Market* **2010,** *16* (4), 297–311.

Von der Dunk, F. G. Space Tourism, Private Spaceflight and the Law: Key Aspects. *Space Policy* **2011,** *27* (3), 146–152. doi:10.1016/j.spacepol.2011.04.015. ISSN 0265-9646.

Wahab, S.; Crampon, L. J.; Rothfield, L. M. *Tourism Marketing: A Destination-Orientated Programme for the Marketing of International Tourism*; Tourism International Press: London, 1976.

Wang, Y.; Krakover, S. Destination Marketing: Competition, Cooperation or Coopetition? *Int. J. Contemp. Hosp. Manage.* **2008,** *20* (2), 126–141.

Wilkie, W. L.; Pessemier, E. A. Issues in Marketing's Use of Multi-Attribute Attitude Models. *J. Market. Res.* **1973,** *10* (4), 428–441.

Wirtz, J.; Lovelock, C.; Chatterjee, J. *Service Marketing*, 8th ed.; Pearson, 2018.

WTM Team Innovative Technology Enhancing Tourist Destinations, 2018. https://hub.wtm.com/innovative-technology-enhancing-tourist-destinations/

WTO. Thinktank. World Tourism Organisation, 2002. http://www.worldtourism.org/education/menu.html

Yoo, K. H.; Gretzel, U. The Role of Information and Communication Technologies (ICTs) in Marketing Tourism Experiences. In *The Handbook of Managing and Marketing Tourism Experiences*; Emerald Group Publishing Limited, 2016; pp 409–428.

Zephoria.com. The Top 20 Valuable Facebook Statistics, Updated August 2020. https://zephoria.com/top-15-valuable-facebook-statistics/

Zhu, Z. Discovering the Influential Users Oriented to Viral Marketing Based on Online Social Networks. *Phys. Statistic. Mech. App.* **2013,** *392* (16), 3459–3469.

360runningbarcelona.com. What's a Running Tour? Frequently Asked Questions|360 RunningBarcelona, n.d. 360runningbarcelona.com

Index

T